MV AGUSTA FOURS

MV Agusta
The Complete Story
Fours

Mick Walker

The Crowood Press

First published in 2000 by
The Crowood Press Ltd
Ramsbury, Marlborough
Wiltshire SN8 2HR

British Library Cataloguing-in-Publication Data
A catalogue record for this book is available from the British
Library.

ISBN 1 86126 329 5

Dedication
To Gary Walker, a much missed son and friend.

Typeface used: New Century Schoolbook.

Typeset and designed by
D & N Publishing
Membury Business Park, Lambourn Woodlands
Hungerford, Berkshire.

Printed and bound by The Bath Press.

Contents

Acknowledgements

With the launch of the new F4 in 1998 MV Agusta was reborn. But would this new machine justify the proud badge? The answer is a positive yes. In fact I would go as far as saying that this is the bike its illustrious marque was born to have. Many, including myself, view it as probably the finest production street bike of the twentieth century – and one which starts the twenty-first century as the motorcycle which many would aspire to own.

But the F4 is not a product of Agusta, but of Cagiva; the Castiglionis purchased the name in the 1980s and to give them credit they wanted a full decade or more until they had a design which justified the name.

But what of the past, the legend of days gone by when MV Agusta ruled the race circuits of the world? Well its all here in *MV Agusta Fours* with the legendary marque's history traced down the decades from its formation at the end of the war, right up to the new era of the F4 and its arrival into the new century as the most lusted after motorcycle in the universe.

But MV is not simply a story of the bikes themselves but the personalities. Engineers like Remor, Magni and Tamburini, riders such as Graham, Surtees, Hocking, Hailwood, Agostini and Read, to say nothing of the autocratic Count Domenico Agusta and

the existing management headed by Claudio Castiglioni.

It is also a technical story with not just the fours, but also the singles, twins, triples and sixes.

John Surtees before his first MV race, Crystal Palace, Easter 1956.

When I wrote *MV Agusta All Production Road and Racing Motorcycles* (Osprey 1987), I was greatly assisted in my task by many MV enthusiasts, this is also the case with *MV Agusta Fours – The Complete Story*. But whereas the earlier title dealt only with the

Giacomo Agostini receiving the laurels of victory, Race of the Year, Mallory Park, 1967.

MV's first star, the Englishman Les Graham, with the 500cc four in the summer of 1951.

production side of the company, this new book sets out to not only bring the story bang up to date, but also include MV's Grand Prix racing successes. And what successes they were too!

Dave Kay and his son Mark provided much encouragement. Dave may well have sparked much controversy over the years, but what no one can truthfully dispute is his genuine love and affection for MV motorcycles.

Its also good to realize that men such as John Surtees, Arturo Magni, Giacomo Agostini and Phil Read are still around and thus bring the history of this great marque to life with their appearances in public at classic events around the world.

Finally my thanks to the various photographers who assisted in my being able to illustrate *MV Agusta Fours*, including Doug Jackson, Roland Brown and Vic Bates in particular. The balance of the photographs came from my own collection.

Mick Walker
Wisbech, Cambs
January 2000

1   The Agusta Dynasty

AGUSTA AVIATION

The MV story really has its beginnings in the year 1907, when wealthy Italian industrialist Giovanni Agusta took the decision to build his own aircraft. This is notable in that it was less than four years after the Wright brothers, Wilbur and Orville, had made their historic first flight by powered machine at Kittyhawk, USA, on 17 December 1903.

Born in the northern town of Parma in 1879, Giovanni Agusta was to play an important role in the fledgling Italian aviation industry. When he began the construction of his own machine – an extremely basic pusher biplane called the Ag1 – he became one of the very first of Italy's aviation pioneers, alongside men such as Alessandro Anzani, Gianni Caproni and Emilio Pensuti. The prototype Agusta Ag1 had its first flight at Capau, just north of Naples, before the end of 1907. It was followed soon after by an improved version, the Ag1 bis.

Another success came in 1911, when Agusta designed one of the first parachutes. His was no ordinary version – it served to save not only the pilot in the event of engine failure, but also the aircraft. Surely it could be seen as a forerunner of a similar device employed today for the recovery of space-rocket capsules.

Also in 1911, Italy became the first nation to carry out a bombing raid, during the conquest of Libya. Following this action, Giovanni Agusta went to Libya as a government volunteer, and served his country in the First World War.

By 1919, Agusta was acknowledged as one the country's leading aviation authorities and the following year saw the formation of Construzione Aeronautiche Giovanni Agusta, a company incorporated for the 'construction, overhaul and repair of both military and civilian aircraft', based in Tripoli and Benghazi. Three years later, in 1923, Giovanni Agusta returned to his Italian homeland and established a new headquarters at Cascina Costa, Gallerate, near Milan.

The new company continued to expand through the 1920s, and in 1927 the Ag2 monoplane had its maiden flight, powered by a 15hp Anzani engine. This had the distinction of being one of the first purpose-built sports planes constructed in Italy.

Just as everything seemed to be progressing so well, on 27 November 1927, less than four years after realizing his dream of creating his own aviation company, Giovanni Agusta died at the early age of 48.

The business passed to Giovanni's wife Giuseppina Turretta Agusta. Giuseppina and Giovanni had had four sons – Count Domenico (the eldest), Mario, Vincenzo, and the youngest, Corrado – and it was Domenico who soon proved to possess a natural flair for the rigours of commercial life. He was very soon put to the test. His father's untimely death was just before the Great Depression, which followed the financial collapse on New York's

Wall Street in October 1929. The young Domenico, then just 21 years of age, came through his test with flying colours, and was to remain at the head of the Agusta empire for many years, until his own death, in February 1971.

PROGRESS

Throughout the 1930s, the Agusta company prospered on the basis of repairs and sub-contracting work, mainly for the civil aviation industry. It also continued to design and build its own aircraft, including the commercially successful AgB6 of 1936, a four-seater, single-engined, light-cabin monoplane.

Towards the end of the decade, Italy embarked on a massive rearmament programme under the dictatorship of Benito Mussolini. In common with the majority of Italian engineering concerns, Agusta either had to accept military contracts, or find itself out of business.

Italy finally joined the war on 10 June 1940, and the Cascina Costa firm was catapulted into the war effort, with rapid growth enforced on both production schedules and the workforce itself. Much of this industrial effort was expended, as before, on sub-contract work. However, Agusta was also given the responsibility for developing its own version of the Breda 88 twin-engined attack bomber, as well as a tricycle landing-gear variant of the Fiat BR20 medium bomber, the first Italian military aircraft to be so equipped.

Like other Italian aviation sector companies, Agusta's real problems began as the war came to an end. Under the terms of surrender, Italian industrial facilities (located mainly in the north, and under German control until the final days) were forbidden to manufacture any products associated with aircraft production.

Fortunately, as early as the autumn of 1943, following Mussolini's deposition, and still under German control, Agusta had begun to consider its post-war plans. In this, the company's darkest hour, Count Domenico came to the rescue. Never one to miss a good commercial opportunity, he had already reasoned that the company's engineering skills were sufficient to undertake virtually anything mechanical. He opted to enter the motorcycle field.

The motorcycle was chosen for several reasons. It was not simply – as many historians would have it – the Count's hobby. The decision was firmly based upon commercial considerations, and the first and sole aim of the MV Agusta motorcycle was to create a profitable enterprise. This was to remain the main aim of the company, at least until the end of the 1950s.

During the Second World War, Agusta was entrusted with several aviation products, including a tricycle undercarriage for the Fiat BR20 medium bomber; the standard version is shown here.

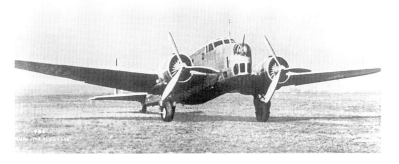

The Agusta family in the early 1950s: Countess Giuseppina Agusta with her four sons (left to right), Corrado (the youngest and the only one still alive), Domenico, Vincenzo and Mario.

A RETURN TO AVIATION

Although Agusta was buoyed up by the success of its diversification into two wheels, Domenico Agusta had not, as some supposed, abandoned the idea of re-establishing the company's aviation division. His efforts to resurrect it were rewarded when, on 4 April 1949, Italy became a signatory to the North Atlantic Pact. With this, companies such as Agusta were able to restart aircraft manufacture.

The new Agusta operation this time concentrated on the design and building of aircraft, in contrast to its pre-war policy of basing its business almost exclusively on repair and sub-contracting. The first fruits of this new direction came with the design of a four-seater cabin monoplane, coded the CP110. This was completed in 1950, and made its first flight the following spring.

Neither the CP110, nor its smaller, cheaper brother, the CP111, met with real commercial success. Instead, the big breakthrough came with Agusta signing an agreement in 1952 with the American company Bell, which allowed Agusta to manufacture the Bell Model 47G helicopter in Italy under licence.

The first production Agusta-Bell 47 left the Cascina Costa factory during 1954. The machine became a best-seller, not just in Italy but throughout Europe. Its influence was such that Bell's chairman was later to comment, 'It really opened up the European market.' The unqualified success of this venture set the scene for future Agusta aviation developments.

There was to be one more abortive attempt at a fixed-wing aircraft, in 1958, in the shape of the four-engined AZ8L. The only kind of its type ever built by Agusta, it proved a flop.

OTHER PROJECTS

After its costly experience with the AZ8L, Agusta concentrated on helicopters, motorcycles and other general engineering work, but there were other projects.

One offshoot of Agusta was SILM, involved in metalwork. Another was a profitable line of diesel engines, both for industrial and automotive use.

Other projects involved three- and four-wheeler cars, including a three-wheel car powered by a four-stroke ohv 349cc twin-cylinder engine, and a number of four-wheel cars designed and produced by OSCA, a company purchased by Agusta in 1960. In 1959, Agusta constructed an experimental hovercraft powered by a 300cc twin-cylinder two-stroke engine. In the 1970s, a tracked military vehicle was designed, which was able to transport personnel or materials over virtually any form of terrain, including deep sand or snow.

Despite these interesting asides, the company's fortunes continued to be based on its helicopters and motorcycles, and the post-war commercial success of the business rested on the strict controls imposed by the Agusta family, notably by Mario and Domenico. Destiny was once again to intervene. First, in 1969, Mario died, followed in early 1971 by Domenico. This left the Agusta group in the hands of Corrado, the youngest of the four brothers.

HELICOPTERS

By the early 1970s, the Italian government had come to regard Agusta primarily as a defence contractor. To protect its interest, it moved in 1973 to take a controlling 51 per cent shareholding. Much of this centred around a plan to develop two new helicopters, the A109 and A129 Mangusta. The success of these two airborne machines was to contribute to the final demise of the original MV Agusta motorcycle company.

The costs of years, not months, of commitment to the development process of a modern helicopter required vast sums of financial backing. It is doubtful if even Domenico Agusta could have fended off government control, given these requirements. Perhaps if he had lived, the two-wheel division might have been allowed to survive. Those who succeeded him were of a different breed.

The original A109A was the first Agusta-designed helicopter to be truly mass-produced and was the end product of a project begun back in 1965. The prototype flew for the first time on 4 August 1971, after Domenico's death, but a crash, protracted trials, modifications and other factors conspired to delay the first production example, which was not completed until 1975. Deliveries began in 1976; it had taken more than a decade to get from drawing board to production line.

The delays were more than offset by the A109's immediate success. Intended originally for commercial use, its design was soon also recognized for its great military appeal. The success of the military A109 was soon followed by the purpose-built A129. Throughout the mid- and late 1970s, Agusta's helicopter business prospered. With the Italian government's financial arm, EFIM (Ente Finanzarin per gli Industrie Metalmeccaniche), taking a controlling interest in the Agusta group, the company became (through the government) the principle shareholders of the long-established SIAM Marchetti (formerly Savoia Marchetti) aircraft corporation in late 1973.

INTERNATIONAL INVOLVEMENT

Following Agusta's expansion, the company took a licence to build Sikorsky's Sea King and Boeing's Chinook helicopters. Under a government directive, Agusta became involved with British helicopter company Westland, and, through a joint venture with Westland, with Aerospatiale of France and MBB of West Germany, in a new project code-named WG34.

The WG34 was first conceived as a detailed study by Westland during 1976–77, financed by a £10 million British Ministry of Defence contract to define a replacement for the Royal Navy's Sea King fleet. At this time, this was the most important rotor-wing aviation venture ever to take place in Europe. In practice, the WG34 project, renamed the EH101, was from 1981 the sole responsibility of the Westland and Agusta companies.

Later, Agusta was one of the same group of companies (plus British Aerospace) that mounted a rescue package to bail out Westland after the British firm hit financial problems at the end of 1985. Their bid, together with another rival bid from Sikorsky/Fiat,

A 1976 advertisement for the new Agusta A109 helicopter.

made headlines in Britain, not only because of the significance of the issue itself, but primarily because of the political storm it created. After Margaret Thatcher's cabinet voted to go for the Sikorsky/Fiat deal, the Trade Minister, Michael Heseltine, walked out of the meeting, telling everyone that he totally disagreed with the decision of his fellow cabinet members.

Rather like Heseltine, however, Agusta staged a comeback. Today, it is an internationally respected organization, employing thousands of workers throughout the world, but it has nothing to do with motorcycles. The Agusta group sold the MV brand name to Cagiva, whilst many of the old factory's racing stock was sold to Team Obsolete during the mid-1980s.

2 On Two Wheels

THE BIRTH OF MV

MV Agusta was officially born on 12 February 1945, when a limited company known as MV (Meccanica Verghera) was incorporated to manufacture and sell motorcycles and associated products. The first steps towards this new venture for the Agusta family had been taken several months before, when Count Domenico Agusta, realizing that Italy might lose the war, began to look at an alternative product range. He had recognized the fact that it would soon be necessary to replace the company's reliance upon the aviation sector.

MV'S 'WASP'

As early as August 1943 – even before the signing of the armistice – the design of an MV motorcycle had begun. The engine of this machine was a simple 98cc (49×52mm) two-stroke with three ports and a two-speed, foot-operated gearbox. Other features included a

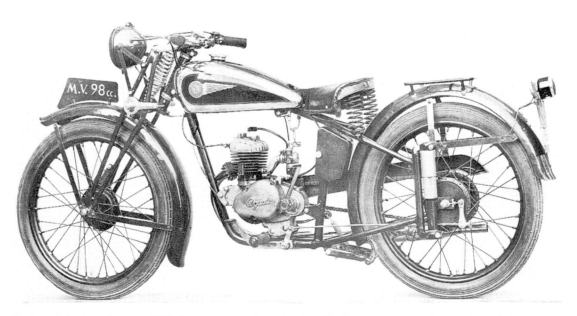

Designed during the war, MV's first motorcycle, a single-cylinder 98cc two-stroke, made its debut in 1945. The original Vespa title was dropped following action by Piaggio, who had already registered the name for its scooter range.

rigid frame, girder front forks and a single, sprung saddle. This buzzing bike was dubbed the Vespa, from the Italian word for 'wasp'.

After some two years of development and testing – much of it under the nose of the Germans, who had remained in the north around Milan until early 1945 – delivery of production bikes began in late 1945. The first dealers to be appointed were Egidio Conficoni and Vincenzo Nencioni, who began to take orders for the tiny MV from premises in the centre of Milan. However, soon after the the bike went on sale, MV was forced to withdraw the Vespa name; the giant Piaggio concern had already registered it for its own range of scooters. So, it was back to 'MV 98', and simply displaying the MV gear-cog emblem on the tank-sides of the bike.

A deluxe model, the 98 Lusso, arrived at the end of 1946. This sported telescopic forks and plunger rear suspension, but still had a two-speed gearbox. A three-speed version followed in 1948. The 98 was to remain in production until 1949 and played a vital role in getting the fledgling motorcycle business of Agusta off the ground.

TWINS AND FOUR-STROKES

In the austere days of the immediate post-war era, twin-cylinder 125s were a rare sight, but MV presented just such a machine at the 1947 Milan Show. The Zefiro, as it was called, had at its heart a 124.6cc (42 × 45mm) two-stroke twin-cylinder engine with iron vertical cylinders and alloy heads. Of unit construction, the power unit featured a single 22mm Dell'Orto carb, wet, multi-plate clutch and four-speed gearbox. The cycle parts owed much to the 98 Lusso, including the full cradle frame, plunger rear end and teles up front. However, although it generated considerable trade interest at the show, the expected production costs were high, and it was never placed in production.

Another MV was developed at the same time as the Zefiro twin, and this one did enter production. It was a 240.2cc (63 × 80mm) ohv four-stroke single of decidedly British appearance. However, the engine was of unit construction, with wet-sump lubrication. Other features of interest

A pair of the first works racing MV 125cc two-strokes at the 194cc Italian GP, which was staged at the Faenza street circuit in Milan. One of the machines, ridden by Franco Bertoni, won its class, giving the marque its first major victory.

included a plunger frame, with a single front down-tube and twin tubes running under the engine assembly, four speeds, gear primary drive, full-width aluminium brake hubs and both the cylinder barrel and head in cast iron. Although it was never built in vast numbers, the four-stroke remained in production from 1947 until 1951.

For a number of years, the 98cc represented the backbone of MV's production. It was also the basis for the company's first racing machine, which appeared in public for the first time at the 1946 Milan Show. It was destined to take part in several races in 1947, but it soon became clear that a full-sized 125 was needed if the company was to be truly competitive. The result came in the shape of a 123.5cc (53×56mm) racer, again, a single-cylinder piston port two-stroke. Its specification included girder front forks, sprung frame and three speeds. There was a 25mm Dell'Orto SSA carb with remote float chamber, flywheel magneto ignition, a light alloy head and cast-iron cylinder, 19in wheels and a 12-litre fuel tank. The maximum speed approached 80mph (130km/h). Ridden by Franco Bertoni, one of the new 125 MVs was victorious over the Milanese Faenza street circuit, in the 1948 Italian GP.

1949

The year 1949 was a significant one for MV, for many reasons. It heralded the beginning of the FIM World Championship racing series (for 125, 250, 350 and 500cc, plus sidecars), and MV built not only an entirely new engine (with the added advantage of four speeds and full unit construction), but also an improved 125 racer using a tuned version of this powerplant, as well as the new 125 TEL roadster and the 125B scooter. Again, both used the new engine, and four speeds. One feature that was carried over from the 1948 125 racer was the 53×56mm bore and stroke dimensions. Both motorcycles now came with an improved swinging-arm rear suspension, although the girder forks were retained.

1950

In 1950, the 125 racer was replaced by a new four-stroke engine (designed by Ing. Piero Remor), but the customer 'over-the-counter' racer remained a two-stroke, as did the roadster and scooter models. Conversely, the new 123.5cc (53×56mm) dohc single racing engine had nothing in common with the production machines. Besides its gear-driven twin-cam cylinder head, other notable features included hairpin valve springs, dry-sump lubrication, small external flywheel, five-speed gearbox, alloy cylinder and head, duplex frame swinging-arm rear suspension, 21in wheels and megaphone exhaust. This twin-cam 125 single was to remain as a factory entry, albeit in updated form, in the World Championship series until as late as 1960, and in the 125cc Italian championships up to 1964. During that time it was changed in several ways, most notably with modifications being made to the suspension and the chassis, and with detailed engine improvements. In 1950, there was no streamlining, but various fairings were fitted later, including dustbin and dolphin types. At one stage there was even an experimental engine with fuel injection. Ignition began with a magneto and ended (from 1959) with a coil.

1950 also saw the debut of the first four-cylinder model, also designed by Remor (*see* Chapters 4–9). At the end of the year, MV signed Englishman Les Graham, its first foreign rider.

There were various designs of MV Agusta two-stroke engines – this one was from a 1950 production racer.

(Below) *This two-stroke engine is from about 1952; the front section of the crankcases house a magneto.*

with a single shock arrangement. Another cost-cutting measure was the absolute minimum of enclosure for the engine and reduced footboard leg-shield panelling. The Ovunque was offered between 1951 and 1954.

1952

The 125cc production roadster motorcycle continued, but was joined in 1952 by a larger-engined version, with a displacement of 153cc (59 × 56mm), sold as the 150 Turismo or Sport. Like the smaller version, several were built in Supersport form for use in long-distance racing events, such as the Milano-Tarinto and the Giro d'Italia (Tour of Italy).

However, the really big news of 1952 was MV's first world road-racing title, with the 125cc series being won by Englishman Cecil Sandford on one of the works double overhead camshaft singles. In the six-round series, Sandford, introduced to MV by Les Graham, won three of the rounds (Isle of Man, Dutch and Ulster) and was runner-up in two (Germany and Spain). Previously, Sandford had

1951

The most notable addition on the production front in 1951 was the frantic development of MV's scooter models. Not only were the C and CSL models improved, but there were also improvements to a CGT, which had debuted in 1950, and the introduction of the brand-new Ovunque (from the Italian for 'everywhere'). MV felt it already had a scooter that served the middle and top end of the market, and the newcomer was aimed at the bargain-basement sector. Although the engine used the same basic two-stroke unit as before, it now came with three instead of four speeds, along with twist-grip change. The suspension layout was also radically simplified,

raced works Velocettes. He rode for MV until 1954, and also raced factory Guzzis, DKWs and FB Mondials, winning the 250cc title in 1957 on the latter marque.

1953

Sadly, MV team leader Les Graham was killed in the Isle of Man TT in June 1953. In commercial terms, however, 1953 was a good year for MV. Following its success in winning the previous year's 125cc world title, MV was flooded with requests for a production version of Sandford's winning machine. MV complied, producing a machine that was a pretty faithful replica, except that it had a single overhead cam and four instead of five speeds. It was built and sold between 1953 and 1956. MV claimed 16bhp at 10,300rpm with a maximum speed of 93mph (150km/h). Dry weight was 165lb (75kg). MV also built a version for the long-distance races with lighting equipment, plus a racing scooter with both the sohc engine and even the full works dohc unit. On the

(Above) *Englishman Les Graham was MV's first foreign signing, joining the team at the end of 1950. He was the very first 500cc World Champion (on an AJS Porcupine twin) in 1949. Graham was a major influence during the development process of the early MV four-stroke racers from 125 to 500cc.*

Graham's first season, 1951, was dogged by a string of mechanical gremlins, typified by his retirement in both his outings (125 and 500cc) in the Italian GP that year. He is seen here at Monza on the smaller mount. Like the four-cylinder machine, this was the work of Ing. Remor. Its 123.5cc (53 × 56mm) dohc engine was housed in a duplex frame with swinging-arm rear suspension. The disc-type rear wheel and handlebar fairing were soon discarded.

latter machine, there was a tiny gap between the steering head of the frame, and small wheels – otherwise, it had every appearance of a pukka racing motorcycle.

At the end of 1952, MV had displayed a brand-new production roadster with a 172.3cc (59.5 × 62mm) sohc four-stroke engine. This featured a very neat chain-

Another Englishman, Cecil Sandford, was signed for the 1952 season, and repaid MV's trust by winning the company's first-ever world title that year (the 125cc). He is seen here after winning the Ultra-Lightweight Isle of Man TT the same year.

The 123.5cc double overhead cam MV engine: note gear train on offside of cylinder, oil pump, front-mounted magneto, hairpin valve springs and exhaust cam-driven rev counter.

driven single-cam unit with integral four-speed gearbox, wet multi-plate clutch and geared primary drive. It entered production at the beginning of 1953 and was subsequently offered in a number of guises, including CSTL (Turismo Lusso), CS (Sport), CSTEL (Turismo Esportazione Lusso), CSS (Supersport – but more commonly known as the Disco Volante, or 'flying saucer'), and the Squalo ('shark'), a racing-only model.

Another important design making its production debut in 1953 was the Pullman. First seen in public at the Brussels Show in

MV's 125cc team at the 1953 Isle of Man TT (left to right): Cecil Sandford (52), Les Graham (60) and Carlo Ubbiali (79). Graham won the race, but was later to lose his life when he crashed the 500cc MV at the bottom of Bray Hill in the Senior event.

Ubbiali testing the 125 works dohc single at Monza in 1953. Note the Earles front forks and riding position to maximize speed on the straights. This posture was abandoned with the advent of fairings.

Produced between 1953 and 1956, the 125 Competizione was essentially a single ohc version of the factory's dohc. Besides the valve gear, the production racer had to make do with four instead of five speeds. Power output at 10bhp was some 25 per cent down on the factory engine.

January 1953, the Pullman was a distinctive machine. Powered by the long-serving 123.5cc piston port two-stroke engine, which had been seen in a number of earlier MV ultra-lightweight motorcycles and scooters, the Pullman was a masterpiece of marketing, successfully filling the gap between the motorcycle and its small-wheeled brother. It combined the best features of both, having, for example, the larger (15in) wheels of the motorcycle and the convenience of the scooter's effective mudguarding and footboards. The Pullman was a great success – it was built in large numbers, and remained in production until 1957, and, flatteringly, it also led to a number of imitation models from rival manufacturers.

1954

The year 1954 was most significant to MV's future. In this year, the overhead valve engine made its first appearance in a production model, and the 125 single and 350 twin models of the late 1970s were to continue

with this basic layout until the very end. The first engine size, in 1954, was 123.6cc (54 × 54mm); larger versions included the 246.6cc (62 × 66mm), 1956; 83.2cc (46.5 × 49mm), 1958; 172.3cc (59.5 × 62mm), also 1958; 150.1cc (59.5 × 54mm), 1959; 231.7cc (69 × 62mm), 1962; 166.3cc (46.5 × 49mm), 1964; twin-cylinder 247cc (53 × 56mm), 1966; and, finally, another twin 348.9cc (63 × 56mm), in 1970. (The dates are the years when the engines entered production rather than when they were first seen in prototype form.)

RACING

Not only had MV produced a new 350 four in 1953 (*see* Chapter 3), but there were also a number of important dates for single and twin-cylinder racers, which flew the MV flag in the smaller-capacity classes.

A 174.5mm (63 × 56mm) dohc appeared in early 1955. This was then developed into the larger 203cc model and, finally, into a 248.2cc (72.6 × 60mm) version.

The ill-fated 348.8cc (62 × 57.8mm) dohc twin, designed by noted Roman engineer Carlo Gianini, was tested as a prototype in 1957. (Earlier, a road-going version, again only a prototype, had debuted back in 1955.)

A 247cc (53 × 56mm) dohc twin, producing 36bhp at 12,000rpm, arrived in 1959 and was raced until MV quit the 250cc Grand Prix arena in 1961.

Other notable machines included the following:

1957		
499.2cc	(48 × 46mm)	Dohc six*
1965		
124.5cc	(54.2 × 54mm)	Disc valve two-stroke single
1965		
343.9cc	(48 × 46mm)	Dohc three*
1966		
497.9cc	(62 × 55mm)	Dohc three*
1969		
348.8cc	(43.3 × 39.5mm)	Dohc six*

*(see Chapter 12)

PRODUCTION ROADSTERS

Towards the end of the 1950s, with the advent of small and affordable motor cars such as the Fiat 500, motorcycle production in Italy nose-dived. Because of its successful

(Above) *This flag logo was produced by the factory at the end of 1955 to celebrate its four manufacturer world titles up to that time – 1952 125cc, 1953 125cc, 1955 125cc and 1955 250cc. (The 1953 125cc rider's title was won by Germany's Werner Haas on an NSU.)*

As well as its racers, many of MV's early production roadsters were two-strokes, like this 1956 125 Super Pullman model.

The ohc 175 CSTL Turismo made its public bow at the Milan Show in late 1952. Its 172.3cc (59.5 × 62mm) unit construction engine featured a four-speed gearbox. This is a 1955 model.

In stark contrast to its mighty four-cylinder 'fire engines', MV's production range also included small commuter motorcycles, scooters and mopeds. One of the latter is seen here at the Milan Show in 1955 (below left). The ciclomotore (cyclemotor), sold between 1955 and 1959, was similar in appearance and function to the massively successful NSU Quickly. The 47.6cc (38 × 42mm) single-cylinder two-stroke featured three speeds and front and rear suspension (below).

return to aviation, via the Bell helicopter licence agreement earlier in the decade, MV was in a much stronger position than the majority of its rivals, who relied on motorcycle, scooter and moped sales. MV was able to continue both with its GP racing programme and with the development of new road-going motorcycles.

MV also developed a moped (built from 1955 to 1959) and even a brand-new scooter, the Chicco. Built between 1960 and 1964, the Chicco was powered by a newly designed 155.6 cc (57 × 61mm) single-cylinder two-stroke with horizontal cylinder. This assembly was of unit construction and had no connection with any previous MV

engine. Another feature was fan-cooling – the Chicco was the only MV two-wheeler to be thus equipped.

During the 1960s, MV also designed a number of interesting machines that were

(Above) For the 1957 season, MV unveiled the new 246.6cc (69 × 66mm) Raid. Later, a 301cc (74 × 70mm) was also offered. Production of both versions ceased in 1961.

MV brochure showing the Chicco scooter. This was built to rival the success achieved by Lambretta and Vespa, but, when it arrived in 1960, the scooter boom was nearly over.

An overhead camshaft 175cc MV roadster single from the late 1950s. Note the 'banana' rear section of frame, a feature of many MV production bikes right through to the early 1970s.

destined to remain as prototypes only. One was the Bik, a 166cc twin-cylinder four-stroke scooter with semi-hydraulic tappets; another was a 146cc twin-cylinder two-stroke motorcycle with steeply inclined cylinders and a five-speed gearbox.

Instead of taking these two machines into production, MV chose to stick with its tried-and-tested formula of pushrod singles and later twins of 250 and 350 engine sizes. These had gained five speeds by the early 1970s and then, in the mid-1970s, a change was made from round to squared-off outer casings and engine finning. However, together with the more exotic (and more expensive) dohc fours, by then the motorcycle side was a largely unwanted facet of the Agusta empire. Its ultimate demise was no

doubt hastened by the death in February 1971 of MV's driving force, Count Domenico Agusta. Soon after, the government-appointed cost account managers moved in, and MV's fate as a bike builder was sealed.

MV EXPORTS

MV Agusta motorcycles were exported to the following countries: Aden (now South Yemen), Algeria, Angola, Argentina, Australia, Austria, Belgium, Brazil, Cambodia, Canada, Ceylon (now Sri Lanka), Chile, Cuba, Denmark, Finland, Formosa, France, Germany, Great Britain, Greece, Guatemala, Holland, Hong Kong, Iran, Ireland, Israel, Japan, Jordan, Liberia, Libya, Malaysia, Malta, Morocco, Nigeria, Norway, New Caledonia, Pakistan, Paraguay, Portugal, Rhodesia (now Zimbabwe), Romania, Saudi Arabia, Somalia, Spain, South Africa, Switzerland, Syria, Thailand, Tunisia, Turkey, Uruguay, Venezuela and Vietnam.

3 Building the Legend

Count Domenico Agusta was above all else a man who got things done. While others talked about what they might do, he actually persevered and did it.

Motorcycle manufacturer MV Agusta was incorporated as a company on 12 February 1945. Its first product, a 98cc two-stroke commuter bike (*see* Chapter 2), was presented towards the end of that year, with deliveries beginning soon afterwards to the fledgling dealer network. A stripped and tuned-down version of the diminutive 98 formed the basis of MV's first racing effort in 1946, and was followed by a larger 125 two-stroke for the 1948 season. An improved 125 two-stroke, with a new engine, appeared in time for 1949, but, like Moto Morini, the MV 'stroker' was unable to match the all-conquering dohc four-stroke FB Mondial single.

GILERA INFLUENCE

Following the lack of success in 1949, Count Domenico decided that MV needed to go the four-stroke route. How was this to be achieved by a company with no experience in the field other than a slow-selling ohv single 250 roadster? The answer turned out to be simple: head-hunt the best from another company. Because he had his sights set on classes other than just the smaller ones, the Count went the whole hog and bought the services of Gilera chief engineer Ing. Piero Remor, and one of his chief race mechanics, Arturo Magni. It was an inspired move. The new team was able to design not only a new 125 twin-cam single, but also – far more importantly in the longer term – no less than a 500 four-cylinder model. The latter went from drawing board to test-bench in just fifteen weeks; this record-breaking achievement was nothing short of phenomenal.

Bore and stroke dimensions of the 1950 MV four were square at 54×54mm, giving a displacement of 494.4cc. Running on a compression ratio of 9.5:1, the dohc four breathed through a pair of 28mm Dell'Orto carburettors (one pair of cylinders, sharing a single remote float chamber). Weighing in dry at 260lb (118kg), the four-speed 1950 MV 500 produced 50bhp at 9,000rpm, giving a maximum road speed of almost 129mph (206km/h), making it faster than either the twin-cylinder AJS Porcupine or the single-cylinder Norton.

So, how was this machine produced so quickly? The fact is that Remor based his engine very closely on that of the 1948/49 Gilera, for which, of course, he had been responsible. Indeed, the 'new' MV was so similar that, in today's litigious society, Gilera would surely have sought some sort of legal redress. Remor was clearly aware that an exact copy would not be acceptable, so he did introduce some features that were different, most notably shaft instead of chain final drive, and torsion-bar suspension, both fore and aft. Another development was gear-change levers

The Gilera Four

Gilera first hit the headlines in April 1937, when Piero Taruffi broke the one-hour record, held by the Scot Jimmy Guthrie (Norton), at 114mph (182.5km/h). Taruffi raised this to 121.33mph (195.26km/h) over a 28-mile (45-km) course comprising a section of the Bergamo-Brescia autostrada. The machine the Italian was riding was a Gilera, a fully enclosed, water-cooled four developed from the earlier Rondine design, itself conceived

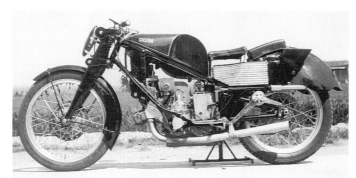

(Above) *A 1935 sketch of the Rondine four, the forerunner of the Gilera multi.*

Water-cooled, super-charged dohc 492.7cc (52 × 58mm) Gilera four, raced by Dorino Serafini to victory in the 1939 German and Ulster GPs.

Ing. Piero Remor (left) with Arcisco Artesiani in 1949, just before quitting Gilera to join MV.

from the even earlier GRB (Gianini, Remor, Bonmartini) air-cooled four-cylinder of the mid-1920s. (Giuseppe Gilera had founded the company bearing his name in Arcore, near Milan in 1911, and had been shrewd enough to spot the Rondine's potential.) The basic design was steadily improved through an intensive racing and records programme, until 1939. Gilera was on course to achieve supremacy in the blue riband 500cc class. Twice during that year, rider Dorini Serafini achieved magnificent victories – beating BMW on its own ground in the German Grand Prix, and winning the Ulster Grand Prix at record speed.

When racing resumed again after the Second World War, Gilera returned with air-cooled four-cylinder machines designed by Ing. Piero Remor (who later left to join MV). At first, their speed advantage was negated by poor handling and lack of low-speed torque, but patient development by Remor's successors and former assistants Franco Passoni and Sandro Colombo eliminated these.

Gilera's first world title came in 1950 (the second year of the official championship series), when Umberto Masetti took the crown. From then on, success followed success, until the Arcore company withdrew from racing at the end of 1957. In eight seasons from 1950, Gilera riders Masetti, Geoff Duke and Libero Liberati

A rare paddock photograph showing both the MV (left, 16) and Gilera (right, 14), Monza, September 1956.

achieved the 500cc championship six times, winning 31 races in the process. It was on a Gilera, too, that Bob McIntyre scored a Junior and Senior TT double, and became the first rider to lap the Isle of Man Mountain circuit at over 100mph (160km/h). In addition, there were the performances of sidecar star Ercole Frigero, constantly brilliant, but

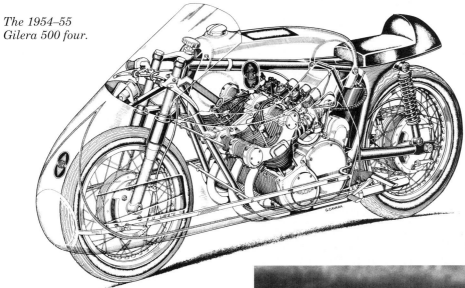

The 1954–55 Gilera 500 four.

sometimes overlooked. For several years he was second only to the great Eric Oliver in the World Sidecar Championship.

When Gilera retired from the sport, it did so on the crest of a magnificent wave, with a series of record-breaking achievements at Monza in 1957, crowned by Bob McIntyre's incredible 141.37 miles (226.19km) in one hour on a 350 four.

Although Gilera made a number of comebacks during the next decade, notably Scuderia Duke in 1963, and the Argentinian Benedicto Caldarella a year later, the company never again recaptured its former glory.

Geoff Duke en route to victory in the 1956 500cc Italian Grand Prix aboard his Gilera four.

on *both* sides of the engine unit. This bizarre innovation required the rider to use his heels, pushing down on the nearside for upward changes, and down on the offside to change down. It was all rather unnecessary, and over-complicated the system. Perhaps these changes to the final drive, suspension and gear change were simply external diversions from the main fact – that Remor had more or less exactly reproduced the Gilera engine.

WORLD CHAMPIONSHIP 1950

The first round of the 1950 World Championship series was the Isle of Man TT; however, although both Gilera and MV entered teams, neither arrived on the island. The MV's debut was at the next round, the Belgian Grand Prix, held over the ultra-fast Spa Francorchamps circuit, in the east of the country in the heavily wooded Ardennes area. MV's four was ridden by another ex-Gilera employee, Arcisco Artesiani, who brought it home fifth, behind the following:

1st   Umberto Masetti (Gilera)
2nd   Nello Pagani (Gilera)
3rd   Ted Frend (AJS)
4th   Carlo Bandirola (Gilera)

Right from its first race, the 1950 model MV four was handicapped by its mediocre handling and lack of road-holding qualities. Journalist Charlie Rous later summed it up as follows:

> The red MV was an ugly brute which certainly did not look inviting to ride. Its high seat and round, bulbous fuel tank did little to blend its form to aid streamlining or provide comfort to assist control.

A week later, at the Dutch TT, Artesiani astonished everybody by out-accelerating the field from the drop of the flag with a surge of power. But it was almost instantly forgotten when he mis-cued a gear change and bent the valves! However, the Italian was backed up by Irishman Reg Armstrong, riding a four-cylinder machine for the first time, who held fifth place until a stop to change a plug dropped him to ninth.

Artesiani was the sole MV entrant in the Swiss GP of Berne, but never featured and was lapped twice by race winner Les Graham (AJS Porcupine). The final classic of 1950 was the Grand Prix of Nations at Monza, which Geoff Duke (Norton) needed to win in order to claim the World Championship – as long as Umberto Masetti (Gilera) finished no higher than third. The outcome was that Duke did win, but Masetti was runner-up and so the Italian won the world title. Perhaps more surprisingly, Artesiani finished a confident third, and MV was on the rostrum in the 500cc class for the very first time. The full result was as follows:

1st   Geoff Duke (Norton)
2nd   Umberto Masetti (Gilera)
3rd   Arcisco Artesiani (MV)
4th   Alfredo Milani (Gilera)
5th   Carlo Bandirola (Gilera)
6th   Dickie Dale (Norton)

At the end of 1950, the four-cylinder MV racing story moved on to its next stage, with the signing of a rider capable of winning the 500cc world title and the Senior TT. In those days, this meant looking for someone from the British Commonwealth – Count Domenico chose to sign 1949 World Champion Les Graham.

A YEAR OF FAILURE

Surprisingly, given Artesiani's 1950 results, 1951 was a year of failure.

From seventeen starts, a 500 MV finished in only four races. The company's best result was a third in the opening round at Barcelona, Spain, when Artesiani came in behind Masetti (Gilera) and Tommy Wood (Norton). The only other leaderboard results

The race team for the coming year, pictured in December 1950 (left to right): four mechanics, Count Domenico Agusta, Ing. Piero Remor, Les Graham, Corrado Agusta, Mario Agusta, Giuseppina Agusta, Franco Bertacchini and Carlo Bandirola. The motorcycle is one of the first 500 MVs constructed.

First outing for MV 'new boy' Les Graham was the Spanish GP in early April 1951. He is seen here (in helmet) looking on as Ing. Remor and mechanic Arturo Magni make adjustments to the 500 four. In the race, after gobbling up places following a poor start, Graham was third on lap 7, only to retire on the next circuit with gear selector trouble.

were gained by new signing Carlo Bandirola, who came fifth in Spain and fourth in Switzerland. Les Graham had a terrible season, failing to score a single point.

For 1951, Graham's machine had been fitted with a telescopic fork, and, for the TT – at which he forced to retire – conventional rear suspension legs replaced the torsion bars.

Les Graham's MV at the 1951 Spanish GP, showing some of the important technical features, including the dual Dell'Orto carbs with separate float chambers, Lucas racing magneto, sand-cast crankcase assembly, exhaust system and the parallelogram (double) torsion bar rear suspension with friction dampers.

(Below) An MV rider trying the riding position of one of the 500cc fours after streamlining had been fitted to the works entries for the 1951 Italian Grand Prix at Monza.

Both team leader Les Graham and Carlo Bandirola (seen here, 50) employed the special fairings, which encased the steering head at the front section of the fuel tank, at the 1951 Italian GP. Bandirola was to be MV's only finisher in the race – coming home ninth, a lap adrift of race winner Alfredo Milani.

(Below) *View of four-cylinder barrel assembly, head valve gear and covers.*

(Below right) *Four carburettors were first tried in 1951; previously, only two carbs were used on the four-cylinder MV.*

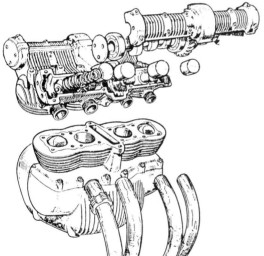

However, the parallelogram (double) swing-ing-arm assembly was retained. Whatever effect the changes had upon the handling were of little use, simply because the bike's mechanical reliability was atrocious.

With the conclusion of such a disastrous season, Remor left. The machine was extensively redesigned in-house for 1952 by

engineers Mario Montoli and Mario Rossi, who worked with chief mechanic Arturo Magni – under the Count's close supervision.

THE 1952 SEASON

Redesign

Most notably, the engine was significantly redesigned, being given the 53 × 56.4mm bore and stroke measurements of the 125

Carlo Bandirola's 500 MV at the 1952 Swiss GP at Berne. He finished third behind the AJS twins of Jack Brett and Bill Doran.

single to produce a new displacement of 497.5cc. With larger valves, hotter camshaft profiles and an increase in compression ratio to 10:1, allied to four carburettors instead of two (first tried mid-way through 1951), power output rose to 56bhp at 10,500rpm. Moreover, an entirely new crankcase, featuring a five-speed gearbox, and chain instead of shaft front drive, made a significant contribution. A new duplex frame, together with a single rear fork, completed the mechanical transformation.

The result, after the eradication of a few early gremlins, was far more encouraging. That year, both telescopic and the new Earles-type front forks were tried.

Les Graham responded by holding second place to Geoff Duke's Norton at the season opener in Switzerland, until being forced to retire with the rear tyre fouling the

Modifications to the 1952 engine also included a five-speed gearbox for the first time on the four. This is Les Graham's machine at the Swiss Grand Prix in April that year.

mudguard. Duke also retired, and the race went to his Norton teamster Jack Brett, from Bill Doran (AJS), Carlo Bandirola (MV) and Nello Pagani (Gilera).

Isle of Man

The second race of the season was the Isle of Man TT, the longest and toughest race in the championship calendar. Graham's MV was the only foreign motorcycle entered in the Senior race, but a combination of poor preparation and bad management prevented an inaugural four-cylinder MV victory.

The favourite, Geoff Duke, led from the off and was an impressive 48 seconds ahead of the field after three laps of the gruelling 37.73-mile (60.5-km) mountain circuit, despite a misfiring Norton engine. Duke refuelled in just 22 seconds, but was then to retire at the end of the fourth lap with clutch trouble.

Graham's pit stop, at 55 seconds, was more than twice as long as Duke's – caused by his overshooting the pit. What should have been a comfortable lead over Reg Armstrong (Norton) was reduced to just 12 seconds. After that, Armstrong got the bit between

his teeth and, by the start of the seventh and final lap, the two were equal. Graham was slowed down on that final circuit when an oil leak soaked his gear pedal and boot, causing his foot to slip. A hashed gear change led to damaged valves when the engine was over-revved, and lost power, and he was also troubled by oil on the rear tyre.

Meanwhile, Armstrong forged ahead and crossed the finishing line to win by 26.6

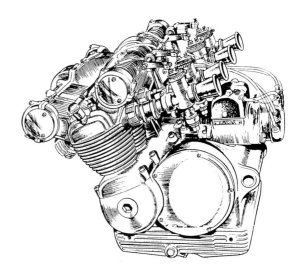

(Above) *Four carburettors were first tried in 1951; previously, only two carbs were used on the four-cylinder MV. The 1952-type engine, much revised from the previous type, included a change from shaft to chain final drive.*

1952 Senior TT. Les Graham with the Earles-forked MV (17) leads the eventual race winner, Norton-mounted Reg Armstrong.

seconds, with an average of 92.97mph (148.75km/h). Extraordinarily, at that exact moment, Armstrong's primary chain broke. If only the race had gone on another few yards, Les Graham, instead of having to settle for second, would have won.

Remaining Rounds

The next two rounds, in Holland and Belgium, did not feature MVs, but Les Graham was back for the German round at Solitude. Here, he finished fourth behind the Norton factory trio of Armstrong, Ken Kavanagh and Syd Lawton, but achieved the distinction of setting the fastest lap.

Victory looked certain at the Ulster Grand Prix in August, when the Englishman had built up a commanding lead. Unfortunately, the infamous bumps of the old Clady Straight brought him to a standstill, with his rear tyre ripped to shreds by a mudguard bolt.

In September, the Italian Grand Prix at last brought Graham victory, rewarding the

(Above) Les Graham (right) and the 1952 125cc world champion Cecil Sandford discussing the Ernie Earles-designed front suspension, used by Graham from mid-1952 onwards on the 500 (and on the 350 when introduced) fours until his death in June 1953.

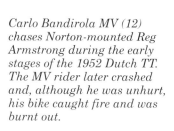

Carlo Bandirola MV (12) chases Norton-mounted Reg Armstrong during the early stages of the 1952 Dutch TT. The MV rider later crashed and, although he was unhurt, his bike caught fire and was burnt out.

New 500 MV four at Monza in early spring 1952. Major changes included engine, chain final drive, plus gearbox and chassis improvements. The tank, fairing and seat were also new.

improvement in speed and mechanical reliability that had occurred over the season. Here, the MV captain won the 125-mile (200-km) race with ease from the Gileras of Masetti and Pagani. Bandirola on another MV was fourth, Colnago (Gilera) fifth, and Armstrong (Norton) sixth.

To win on home ground at Monza was Count Domenico's moment of glory, and he savoured it all the more when Gilera protested, forcing the officials to strip and measure Graham's machine for its displacement size – which proved spot on.

From the speed of Monza, Graham and the team journeyed west to Barcelona, and the twists and turns of Montjuic Park. This was the last meeting of the season, and presented Graham with a real possibility of winning the title, if Masetti's Gilera did not finish. In the event, even though Graham won the race, Masetti finished second, and the championship was his for a second time. Graham was just three points behind.

DASHED HOPES

After this success of 1952 – as well as second place in the 500s, MV had won the 125cc championship, with Cecil Sandford in the saddle – the Count believed that 1953 would see his much-improved machines dominating their respective classes. Les Graham raised his hopes early on, taking one of the 125 dohc singles to victory in the Ultra-Lightweight TT on the Isle of Man, in the first round in the championship series.

However, 'Lady Luck' was about to dash the hopes of the Count and his team in the cruellest way. At the beginning of the second lap of the Senior TT, Graham crashed heavily at the foot of Bay Hill and was fatally injured.

Graham had fallen during practice and this had probably left him in a weakened state, but the official reason for the accident was that a retaining bolt for one of the front suspension legs of the Earles forks had come loose, allowing a leg to go adrift, with complete loss of control.

Les Graham with his works four-cylinder MV in the 500cc race at Ferrara, spring 1953.

Count Domenico was personally so shocked and saddened by Graham's death that the fours hardly made an appearance until the season's end. In mid-season, at Schothen, Carlo Bandirola gave the new 349cc (47.5 × 49.3mm) four its first victory, after Graham had debuted the machine in the Junior TT a couple of days before his fatal accident. Unfortunately, once again the gods were not smiling upon MV. Several of the leading riders had refused to compete in the interests of safety, and the FIM subsequently

Leslie Graham – MV's First Foreign Signing

Although later MV riders might be better remembered today, it was Englishman Leslie Graham who helped more than most to develop the four-cylinder MV into a world-beating motorcycle.

Born on 14 September 1911, in Wallasey, Merseyside, Robert Leslie Graham spent what would surely have been his greatest racing years in the service of his king and country. He won the DFC for his war-time exploits as an Royal Air Force pilot, but he was also a skilled engineer and, perhaps most importantly in his post-war racing career, a man of considerable bravery and skill, and a born leader.

The first of the FIM's World Championship series was staged in 1949 and Graham, then 37 years of age, was the first holder of the blue riband 500cc title on a works AJS Porcupine dohc twin, with victories in Switzerland and Ulster, plus a runner-up spot (behind his main rival Nello Pagani on a Gilera four). In the following year, he came third in both the 500c and 350cc championships, again on AJS machinery.

His record spoke for itself and it was no surprise that Count Domenico Agusta should make Graham his company's first foreign signing at the end of 1950. During 1951, the Englishman rode both the fast ill-handling 500 four and the dohc 125 single, but success did not come until 1952, when he achieved a second place behind the Gilera-Masetti combination.

The first round of the 1953 championships was the Isle of Man TT. Graham had won the 125cc event, and many expected that he would enjoy victory in the Senior event a few days later, and also have world-title prospects. However, the hopes were to end instead in tragedy, when MV's team leader lost control of the Earles-forked 500 MV four at the bottom of Bray Hill. In the ensuing crash, he was killed instantly. Following his death, MV abandoned the bigger classes until almost the end of that season, and development of the four did not really get back on track until John Surtees joined the team in 1956.

Les Graham was survived by his wife Edna and sons Stuart and Christopher. Stuart Graham went on to become a well-known racer, riding works bikes for Honda and Suzuki.

500 Four (1953)

Engine

Type	Air-cooled dohc four, across-the-frame
Bore and stroke	53×56.4mm
Capacity	497.5cc
Compression ratio	10:1
Carburation	Four Dell'Orto 28mm SS carburettors with remotely mounted float chambers
Lubrication	Wet sump
Max. power	56bhp @ 10,500rpm
Fuel tank capacity	5.28 gallons (24 litres)

Transmission

Gearbox	Five speeds
Clutch	Wet, multi-plate
Primary drive	Gears
Final drive	Chain
Ignition	Lucas magneto

Frame

Tubular steel, closed duplex cradle

Suspension and steering

Suspension	front	Earles fork
	rear	Swinging arm with teledraulic dampers
Tyres	front	3.00×19
	rear	3.50×18

Brakes

Full-width aluminium drums, front and rear

Dimensions

Dry weight	260lb (118kg)

Performance

Top speed	125mph (201km/h)

announced that the results would not count towards championship points.

In the months after Graham's death, MV gave a number of riders 'one-off' tests, in an attempt to find a suitable replacement for its late team leader. These included the Englishman Bill Lomas and the German Hans Peter Müller. For 1954, MV signed Lomas and Dickie Dale, and also provided bikes for Carlo Bandirola and Nello Pagani. Pagani, now coming to the end of his riding career, was mainly involved in testing.

1954–55

Further Technical Improvements

On the technical front, the engine now yielded almost 60bhp, although the changes represented attention to detail, rather than major technical development. The two major areas of work concerned the Earles front forks, which now featured straight instead of curved arms, and streamlining. MV used several types of fairing, from a

The 350 Four

Making its track debut in June 1953, in the hands of Les Graham at the Junior TT, the 350 four was first conceived as a training vehicle, rather than as a serious racer with which to score wins. At the time, the 350cc category was far more popular in Great Britain and the Commonwealth than in Italy.

The first 350 engines were built by simply sleeving down the cylinders of the 500 four. Only later, when the motorcycle began to do well in its own right, did MV manufacture a purpose-built 350 engine, following the basic layout of its larger brother (in the same way that British 350s were built).

The original prototype engine used by Graham during the 1953 TT – mainly to provide the Englishman with additional practice laps – produced little more power than its single-cylinder rivals such as the dohc Norton, AJS 7R or Velocette KTT MK8, or, for that matter, the new Moto Guzzi single. On the debit side, it weighed virtually the same as the larger four-cylinder MV.

In its first season, 1953, it took part in only two races that counted towards the World Championships – the TT and, later, the German GP, where it won over the extremely testing and dangerous Schotten circuit. During the rest of the year it was used only for testing purposes.

In the following seasons, before being outpaced by the more modern four-cylinder Hondas (from 1962 onwards), the 350 four made a valuable contribution to providing double-capacity track time for the various MV riders, including Surtees, Hocking and, finally, Hailwood. It won the 350cc world title in 1958, 1959, 1960 and 1961.

Persistent clutch slip slowed the progress of the new 350 MV four when it made its debut during the 1953 Junior TT. Rider Les Graham was forced to retire after two comparatively slow laps. Later the same week, Graham died when he lost control of the larger MV four during the Senior race, and was killed instantly.

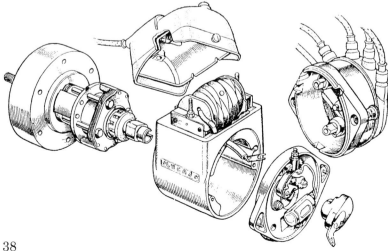

Lucas four-cylinder rotating magnet magneto, as fitted to the 1953 MV multi.

After Graham's death, MV was not represented in the larger classes until Carlo Bandirola rode a 500cc model to runner-up position during the final round of the series in Spain during October. Dickie Dale (seen here testing one of the fours) joined the team for 1954.

(Below) The magnificent MV Agusta Trophy (showing Nello Pagani on the 500/4) was presented by the factory to the organizers of the Milano-Taranto races. It was won outright by Laverda during the mid-1950s.

Nello Pagani acted as team manager and chief tester for much of the mid- and late 1950s. He is seen here during 1954 on one of the 500 fours.

simple handlebar mounting device, which was little more than a front number-plate holder, to a fully enclosed dustbin type, by now beginning to be favoured by the likes of Gilera, Moto Guzzi, FB Mondial and NSU. MV was at first reluctant to put a full 'dustbin' on the 500, because the machine, being quite tall and heavy, was easier to control

without streamlining. In the end, however, the company had to go with the flow – handing such a speed advantage to the opposition was unthinkable.

As for results, Bandirola (by now almost 40 years of age) was the most consistent, but it was Dale who not only won MV its sole 500cc victory (in the first round in Spain) during 1954, but also finished the season fifth in the 500cc championship table.

A 'Star' Name

Generally, 1954 was a poor year for MV, made even worse by NSU's complete dominance of the 125/250cc categories. Count Domenico decided that another big signing was needed.

Dickie Dale rode this part-streamlined MV four in the 1954 Senior TT. He finished seventh at an average speed of 83.14mph (133km/h).

(Below) *Dickie Dale at Signpost Corner during his 1954 Junior TT ride on the smaller four cylinder MV.*

In contrast to Dale's machine, this MV 500 raced by Bill Lomas in the 1954 Senior TT had only a flyscreen for the rider's protection. Lomas lay seventh at the end of lap 3 – two places ahead of team-mate Dale. At the end of lap 4 the race was sensationally stopped, due to appalling weather conditions, but Lomas retired on that lap, so was classed as a non-finisher.

(Right) *This is just one of several fairings race-tested by MV during the mid-1950s. Carlo Bandirola is seen here during the Italian Senior Championships at Senigallia, August 1954.*

The man chosen was the Rhodesian Norton star, Ray Amm, who had finished the 1954 season runner-up in both the 350 and 500cc title chases. At the same time, both Dale and Lomas quit, going to ride for Moto Guzzi, but deals with other new recruits more than offset these losses, with Umberto Masetti, Tito Forconi, Luigi Taveri and Remo

(Left) *The 350 MV four after the fatal accident that claimed the life of the former Norton star Ray Amm at Imola on Easter Monday, 11 April 1955.*

Count Agusta had seen the Rhodesian Ray Amm, pictured below, as a natural successor to the late Les Graham. Amm's death was to be almost as big a blow to MV as that of Graham two years earlier. It was only with the signing of John Surtees, almost a year later, that the team was to stage a recovery.

Venturi all signing on the dotted line. With the existing team members of Bandirola, Pagani, Coppeta and Ubbiali, MV started the 1955 season with no less than nine riders.

With NSU in retirement, the appearance of a brand-new 203cc (68 × 56mm) dohc single for the 250cc GP class, and a 174.5cc (63 × 56mm) version for the national Italian Formula 2 events, and improvements to the four-cylinder 350 and 500cc models, MV appeared well equipped to mount a serious offensive in all four solo capacity classes counting towards the World Championship series.

Disappointment and Tragedy

In fact, only Carlo Ubbiali's 125cc championship was to come home to Cascina Costa. The year was to prove to be a mixture of disappointment and tragedy.

Tragically, Ray Amm was killed in his very first race for the company, riding the 350 four at the Imola Gold Cup on Easter Monday. This was a major blow – the Count had seen Amm as Graham's natural successor

and MV's new team leader. Another blow came when Bill Lomas resigned to ride the new 203cc single. He was denied the championship after being excluded following his win in the Dutch TT; it was alleged that, by keeping his engine running, he had failed to comply with the rules on refuelling.

The Count might have been tempted to give it all up and save his money, but he chose not to. And he found not only a new leader, but also a 500cc MV world champion. That man was John Surtees.

4   Grand Prix Success

SURTEES SIGNS

After the untimely death of first Leslie Graham, and then Ray Amm, and with Gilera dominant in the 500cc class of the World Championships – Geoff Duke had won the title in 1953, 1954 and 1955 – Count Domenico Agusta needed to come up with something special. His answer was to sign the up-and-coming English rider, John Surtees.

As proficient in the workshop as he was on the track, Surtees' signing marked a turning point in MV's history. He was, more than any other rider, responsible for bringing the 500 MV to its definitive form. After his retirement at the end of the 1960 season, it was 'his' machines that Hocking and then Hailwood raced so superbly. They were withdrawn only after 1965, in favour of the new triple (*see* Chapter 12).

Not since Les Graham had the MV engineers had someone who was so good at pinpointing faults on a machine. Surtees had an engineering background, and was also a brilliant rider, gifted with a combination of smoothness and control. This meant that he was able to achieve superior lap times while others might have been far more ragged and thus more accident-prone.

TECHNICAL DEVELOPMENTS

Although 1955 had not been very successful from a results point of view in the 500cc division, MV had carried out considerable technical development. Streamlining again took up a significant amount of time. There were full 'dustbin' fairings, with and without cooling airscoops for the engine and brakes, and even 'half dustbins' were race-tested. Eventually, it was decided to use a combination of a dustbin fairing with a cooling scoop on either side of the front wheel, with a streamlined nose cone, which not only featured a perspex screen, but also a base that extended backwards for over half the tank's length.

The frame was redesigned to give a lower centre of gravity, and a lower riding position, while a detachable cradle was incorporated in the interest of simpler engine removal. The Earles front fork was formally abandoned in favour of a new MV-built telescopic front fork of the leading-axle type, which featured multi-role coil springs above the alloy sliders. Various rear shock absorbers were tried (MV and British Girling types); the double-sided front brake was equipped with massive airscoops – on an aircraft they would have been called airbrakes! The fuel tank was also reshaped to provide a longer, lower style.

With attention to detail and minor modifications, the power had been increased to 65bhp at 11,000rpm. For 1956, this was increased to 67bhp at the same engine revolutions. Again, the brakes were modified, this time with a number of small holes to improve internal air circulation. A form of ram-air induction was also introduced, with a pair of tubes, one each side, protruding from the nose of the fairing to ram air into the carburettors. Further development work went into finding the most suitable dustbin fairing, but otherwise the 1956 machines were much as they had been in 1955.

Finally, another important development had taken place, this time within the organization, with the appointment of former rider Nello Pagani as team manager. Although Count Domenico's hand remained firmly on the tiller, Pagani did have some influence. He managed to make a valuable contribution to the team's overall efforts over the coming years without coming into conflict with the autocratic Count. Agusta was not an easy man to work with – in fact, a number of observers have gone so far as to describe him as a despot, although not to his face!

SURTEES' FIRST SEASON

The Surtees/MV combination first appeared in Britain at the south-east London Crystal Palace circuit on Easter Monday, 2 April

1956. The machine Surtees rode to victory that day was actually a 1955 machine, as the very latest bikes were kept back for World Championship duties.

The six-round 500cc world title chase got under way in the Isle of Man in early June with the Senior TT. MV's chances were assisted by the fact that reigning champion Geoff Duke, together with his Gilera team-mate Reg Armstrong, had been suspended by the FIM until 1 July for having supported a private riders' strike at the 1955 Dutch TT.

John Surtees did not have it easy, however. Even though he won the Senior TT, he did not do so on his new 1956 bike. During the final practice session, he collided with a cow that had wandered on to the Mountain Road section of the course. The cow was killed and the bike was too badly damaged to be repaired in time for the race. Fortunately uninjured, Surtees raced and won on a 1955 model with a half fairing fitted. The final result was as follows:

1st John Surtees (MV
2nd John Hartle (Norton)

The first British appearance of the 1956 MV 500 four, with its new rider John Surtees, was at London's Crystal Palace circuit on Easter Monday, 2 April that year. Although the engine was similar to before, it had been modified to give increased power, now almost 70bhp at 10,500rpm. There was also a new frame, together with new front forks of the leading-axle type.

Surtees got off to a cracking start in the 1956 500cc world championship series by winning the Senior TT.

(Below) *MV's John Surtees (centre), flanked by second-place John Hartle and third-place Jack Brett (left and right respectively, both Norton-mounted), after the 1956 Senior TT.*

3rd   Jack Brett (Norton)
4th   Walter Zeller (BMW)
5th   Bill Lomas (Moto Guzzi)
6th   Derek Ennett (Matchless)

Surtees went on to win the Dutch TT at Assen (with Walter Zeller finishing second on his works BMW).

The first Surtees–Duke clash of the season came at the Belgian Grand Prix, the fastest race on the calendar at the time, held over the famous 8.8-mile Spa Francorchamps course in the heart of the Ardennes. Duke rocketed ahead and shattered the lap record. Then, on lap 13, when leading by almost a minute, his Gilera expired with piston problems.

After this, Surtees was left to win at his own pace, to take his third victory in a row. Effectively, this meant that he could not be beaten in the championship. As events were to prove, he was to need this Belgian result; in the very next round, at Solitude in West Germany, he crashed in the 350cc on the smaller four and broke his arm. This injury ruled him out of the remaining rounds of the championship, which were taken by three different winners: Reg Armstrong (Gilera) in West Germany; John Hartle (Norton) in Ulster; and Geoff Duke (Gilera) in Italy.

As a result, the 1956 500cc world championship went to John Surtees of MV Agusta. Carlo Ubbiali won both the 125 and 250cc titles on MV singles, and it was a year to remember for the Cascina Costa marque.

WORK FOR THE 1957 SEASON

Even though much of the work was not associated with the four-cylinder model, 1956 was nevertheless a year of intense technical development at MV. The factory had three types of 350cc class machines on the stocks at the time: the four, plus a twin and a single. It was also experimenting with features such as fuel injection, four valves per cylinder and desmodromic (positive) valve operation. Another project was a 125cc dohc single with a box-section, light-alloy riveted frame. The top section of the frame served as a fuel and oil container, while the engine-gearbox unit completed the frame at the bottom.

The most glamorous project of all was a top-secret 499.2cc (48 × 46mm) six-cylinder model (*see* Chapter 12), which was not revealed to the public until several months later, towards the end of 1957. This MV six-cylinder was being developed for a simple reason – not only was the Gilera four still a major force, but also Moto Guzzi had brought out a V-8, which seemed capable of making the fours obsolete.

As for the *quattro*, the 500 gained new bore and stroke dimensions of 52 × 58mm, giving a displacement of 492cc, while the frame was again lowered.

Looking back, the 1957 season was probably the pinnacle of Grand Prix racing of that decade – indeed, some would consider, of any decade. Gilera fielded not only Geoff Duke, but also top men such as Bob McIntyre and Libero Liberati. Moto Guzzi had a vast team, which included Bill Lomas (although an accident at the Imola Gold Cup in April 1957 kept him out of action), Keith Campbell and Dickie Dale. BMW had Walter Zeller, and MV had John Surtees, Umberto Masetti and new signing Terry Shepherd. In the six-round 500cc series, Libero Liberati came out top with three victories (Hockenheim, Ulster and Monza); Bob McIntyre won the Senior TT (as well as the Junior event for good measure in what was the 50th Golden Jubilee of the TT series); and Norton-mounted Jack Brett won a sensational Belgian GP in which

Start of the 1957 500cc Italian Grand Prix at Monza. MVs (2, 60 and 36), Gileras, BMWs, Nortons and Moto Guzzis – what a line-up!

there were only five finishers, with Liberati being excluded for using a machine from another team member (Bob Brown). Finally, John Surtees won in the Dutch TT at Assen, beating Liberati and Zeller.

The final championship table read:

1st   Libero Liberati (Gilera)
2nd   Bob McIntyre (Gilera)
3rd   John Surtees (MV)
4th   Jack Brett (Norton)
5th   Geoff Duke (Gilera)

THE END OF AN ERA

At the end of September 1957, and only a few days after the final round of the championship series at Monza, the shattering news broke that Gilera, Guzzi and FB Mondial would not be racing in 1958. Dr Gerado Bonelli, Director General of Moto Guzzi, made the official announcement at a press banquet (later dubbed the 'Last Supper' by Italian journalists) in the presence of representatives from Gilera and FB Mondial. Various reasons were given, including the fact that all three marques wanted to 'demonstrate the undeniable technical excellence of their products' and that 'recently there had been no foreign opposition'. In addition, there had been an outbreak in Italy of anti-racing public opinion, fuelled by the popular press and the restrictions of the authorities following the Mille Miglia tragedy earlier that year, when several spectators had been killed.

The *real* reason, which obviously could not be given at the time, was that motorcycling all across Europe was in trouble. It was the beginning of an era during which the public would turn away from motorcycles to small cars in ever-increasing numbers.

Although MV's main rivals (even BMW's star rider Walter Zeller had quit) had left the scene, the World Championship series did not change. MV dominated every solo class until the end of 1960, but that does not mean that they didn't have to try hard, or that they were not competitively driven. There were still races to win and records to be beaten – the latter made more difficult by the FIM's decision to ban dustbin fairings at the end of 1957.

Despite the lack of credible opposition, and the fact that for the 1958 season MV had to employ less streamlined dolphin fairings, Surtees still managed to set new lap records at both Assen and Spa Francorchamps. In the following year, 1959, he smashed not only the Isle of Man lap record, but the Monza one as well.

By now, the 500 four's engine was developed to almost 70bhp, while the smaller 350 four was able to achieve 45bhp. There were also new MV-manufactured four-leading-shoe front brakes.

A GOLDEN AGE

Although development of the six continued, it was now much less of a priority. In fact, its only race was destined to be the Grand Prix of Nations at Monza in September 1958. With John Hartle at the controls, it was forced to retire at mid-distance, after suffering a poor start.

Hartle had been signed to replace Terry Shepherd, who, although an excellent rider, never seemed to reach his full potential on the MV fours. Like Shepherd, Hartle had come to the Count's attention after a string of superb performances on single-cylinder Nortons, including a win in the 1956 Ulster GP.

The years 1958, 1959 and 1960 are now considered by many to be the golden age of MV. During that period, out of seventy-six races counting towards the World Championships (in 125, 250, 350 and 500cc classes), and without always taking part (for example, after championships had been won), the factory was still victorious in no less than sixty-three. With thirty-two victories, Surtees was by far the most successful (even though he only took part in the 350 and 500cc classes), followed by lightweight jockey Carlo Ubbiali, with seventeen wins.

Besides its success in the world series, MV was also dominant in the domestic Italian championships, with Carlo Bandirolo, Gilberto Milano and Tino Brambilla all winning titles.

There were more technical changes in 1959. On the 500 these were relatively minor, the most significant being a modification to the frame to give a combination of maximum rigidity and ease of maintenance. The angle of inclination for the rear shock absorbers (now usually of Girling origin) was altered several times during this season. There was a long-distance (40-litre) fuel tank expressly for the Isle of Man, while the front-brake airscoops were also modified.

More work, however, was put into the smaller four. In an attempt to improve its power-to-weight ratio, special materials from Agusta's aviation division were employed,

reducing the weight by 45lb (20kg). Then, in 1960, a new version of the 350 appeared with a larger-capacity sump, improved streamlining, and a space-type rear-frame assembly.

(Above) *John Surtees pushes his 350 MV four on which he won the 1958 Junior TT. Also in the picture is Bill Webster (cap) and team manager Nello Pagani (far right).*

First lap photograph of John Surtees at Union Mills on his way to winning the 1958 Junior TT at an average speed of 93.97mph (151.3km/h)

After his excellent performances on private Nortons, John Hartle was signed up as support rider to Surtees for the 1958 season. He is seen here (33) at the Ulster Grand Prix, where he gained second in the 350cc and third in the 500cc races. Team manager Pagani is to Hartle's right.

John Hartle (4) leads John Surtees during the 1958 350cc Italian GP at Monza. Surtees had taken the lead by the end of the race, with Hartle second.

(Below) Surtees winning at Aintree near Liverpool, September 1958. The end-of-season British short-circuit meetings usually saw the top stars once the world series had been concluded.

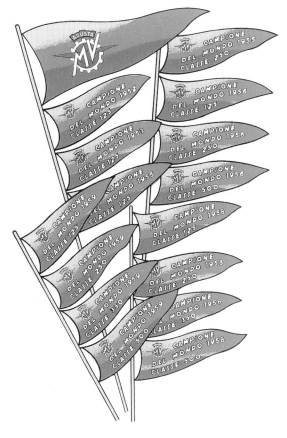

(Above left) *John Surtees rounds Governors Bridge on his way to victory in the 1959 Junior TT.*

(Above) *The decal that adorned the tank tops of production MVs from the end of 1959 – it recalled the factory's superb record on the race circuits of Europe throughout the decade. It was an achievement that was unrivalled.*

Surtees and Hartle with their 350 MV fours after finishing first and second respectively in the 1959 Junior TT. Norton-mounted Scot Alastair King is on the left.

Surtees' team-mate Hartle with his 350 MV four during the 1959 Junior TT.

(Below) *Surtees studies one of the MV fours he rode to a double victory in the 1959 350 and 500cc Ulster GPs held over the Dundrod road circuit on the hills overlooking Belfast. Unchallenged, he averaged 91.32mph (146km/h) in the smaller capacity category and 95.28mph (152.5km/h) for the blue riband 500cc event.*

Also in 1960, MV had new signings in the shape of Remo Venturi, Gary Hocking and Emilio Mendogni. John Hartle's contract was not renewed at the end of 1959.

Major changes occurred at the end of 1960. John Surtees and Carlo Ubbiali both chose to hang up their leathers, and the Count responded by reorganizing the team. 'Private MV' stickers were applied to the machines' fuel tanks and only Hocking was entered for the 1961 World Championship series. At the start of the season, he took

At the Italian Grand Prix at Monza on Sunday 6 September 1959, John Surtees scored yet another 350/500cc double for MV. Second place in both races went to another MV rider, Remo Venturi (to Surtees' right).

For 1960, MV had three riders in the larger classes: Surtees, Venturi and newcomer Gary Hocking. At first, only Surtees and Venturi were allowed on the 500s. This photograph shows the pair on the start line (Venturi nearest the camera) of the French GP. Surtees won with Venturi second.

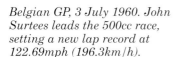

Belgian GP, 3 July 1960. John Surtees leads the 500cc race, setting a new lap record at 122.69mph (196.3km/h).

part in the 250, 350 and 500cc classes, but early races proved that the 250 twin was no longer competitive against the latest Japanese machinery (notably the four-cylinder Hondas), and MV quit the class. For the Italian national championship, Remo Venturi and Silvio Grassetti were contracted.

MV retired in the TT and did not attend the final round in Argentina, but won the remaining eight rounds of the 1961 500cc World Championships in West Germany, France, Holland, Belgium, East Germany, Ulster, Italy and Sweden. Gary Hocking won seven of these, the eighth went to new signing Mike Hailwood on his first entry for MV at Monza.

In the 350 category, Hocking won four of the seven rounds (to add to his season-opening 250cc victory in Spain).

John Surtees, Champion on Two and Four Wheels

Born into a family of motorcycle enthusiasts in Catford, south-east London, on 11 February 1934, John Surtees has the distinction of being the only man ever to have won both a two-wheel and a four-wheel racing world championship title. He was also the first man to win a world crown on one of the four-cylinder MV models. In all, Surtees won seven world titles on MVs – four 500cc, three 350cc – between 1956 and his retirement, at the end of 1960. His solitary four-wheel championship was the 1964 Formula 1 title, which he took with Ferrari. Later still he drove for Honda, before becoming a team manager.

Surtees' father Jack was a successful competitor himself, most notably with an HRD Vincent-powered sidecar. John's career began with a Vincent Grey Flash, while he was employed at the Stevenage factory in 1950–51. By 1952, he was racing Nortons, eventually becoming a works-supported Norton rider for the 1955 season. His performance, including beating Geoff Duke and the four-cylinder Gilera, brought him to the attention of Count Domenico Agusta, who signed him to race both the 350 and 500cc four-cylinder MVs in 1956.

Even though Surtees' main rival Duke was barred from the first half of the season for supporting a riders' strike at the Dutch TT the previous year, nothing can detract from the fact that, in his first year as an MV rider, Surtees won the company its first-ever 500cc world title. This was achieved even after Surtees put himself out of action at the West German GP at Solitude after falling and breaking his arm on the 350 four.

MV didn't win a single championship in 1957, but, after the withdrawal of Gilera, Moto Guzzi and FB Mondial, got back on track to win virtually everything until the Japanese arrived.

When compared with all the other MV champions, Surtees ranks alongside the very best. And, as anyone who has seen him ride in modern classic events will testify, John Surtees certainly has lost nothing of his track craft.

HOCKING AND HAILWOOD

Mike Hailwood had been signed by MV following a long list of major achievements, culminating in no less than three TT victories in a week during June 1961, and the 1961 250cc World Championship title (on a Honda). His first appearance for MV was at Monza for the penultimate round of the series, and he won the 500cc, while Hocking took the 350cc.

With two racers of such ability, many racegoers were anticipating the 1962 championship with relish, and particularly the first round, the Isle of Man TT. They were not to be disappointed.

Hailwood proceeded to take victory from his team-mate in the Junior TT (350cc) event, while Hocking got his revenge by winning the Senior TT a couple of days later. However, on the very evening of his triumph, Hocking stunned the racing world with the announcement that he was to quit motorcycle racing. His explanation was that the death of his friend, the Australian Honda rider Tom Phillis, who had ridden a 285cc four-cylinder model in the Junior TT, had forced a decision that he had already been considering. Hocking then went racing on four wheels, only to suffer a fatal accident at the wheel of a Formula 1 car in South Africa during December 1962.

Gary Hocking's withdrawal from the team after the Senior TT meant that, for the remainder of the 1962 season, Mike Hailwood was MV's sole representative in the World Championship. He responded by winning five 500cc GPs on the trot, at Assen, Spa Francorchamps, Dundrod, Sachsenring and Monza.

Gary Hocking – The Forgotten Champion

In the list of MV's world champions, Surtees, Hailwood, Agostini, Read and Ubbiali immediately spring to mind. Two important names are often forgotten – Cecil Sandford, who won the marque's first-ever world title (the 125cc) in 1952, and Gary Hocking, the brilliant Rhodesian.

Hocking reigned between Surtees and Hailwood and was double 350/500cc title winner in 1961. One reason why his Grand Prix career is less well remembered than others is because it lasted just four years from novice to double world champion. Another is because, unlike Surtees and Hailwood, he very rarely raced on the British short circuits. Yet Hocking, who tragically died in a racing-car accident after quitting motorcycle racing in December 1962, gained a distinction achieved by neither Surtees nor Hailwood: he became a full works rider in only his second GP year.

On his way to becoming double MV World Champion in 1961, Gary Hocking rode Norton singles for Manchester dealer Reg Dearden in the late 1950s, followed by a factory ride with MZ during the first half of the 1960 season before being signed by the Count.

As the late Charlie Rous once said, 'Welsh-born Hocking was quiet and did not waste words. But he was as tough as granite, and his seemingly shy manner hid extreme self-confidence and determination. With only limited racing experience, he firmly believed that he could beat anybody.'

Gary Hocking was taken to Rhodesia (now Zimbabwe) by his parents as a nine-year-old in 1946, and stayed on in Africa when the family returned to the UK in 1955. His first motorcycle was a twin-port Jawa two-stroke, then came a couple of Triumphs, including a new Tiger 110. A friend from his schooldays, veteran race mechanic Nobby Clark, remembers the young Hocking winning seventeen consecutive races on a dirt track at Umgusa on one of the Triumphs.

His first outing on a pukka race bike was on a 350cc Manx Norton, and this was followed by a stint in partnership with future Honda multiple world champion Jim Redman. In March 1958, Redman left to race in Europe and Hocking found himself unable to continue the business on his own. He followed Redman to Europe and was fortunate enough to receive backing from Manchester dealer Reg Dearden with a couple of Manx Nortons. In his first-ever race for Dearden, in the Dutch TT at Assen, Hocking stunned the racing world by finishing a magnificent sixth in the 500cc, behind the MV fours of John Surtees and John Hartle, Derek Minter's Norton and the BMWs of Hiller and Dale. He went even better at the West German GP over the infamous 14.7-mile Nürburgring, where he finished third behind Surtees and Hartle.

Dearden's support continued into 1959 and then, after some impressive early-season results, including the non-championship Austrian and Saar GPs, Hocking was eventually signed by the East German MZ squad.

In his first race on the air-cooled MZ 250 twin in the Swedish GP, he beat MV world champion Carlo Ubbiali by over a minute, and then went on to win the 250cc Ulster GP. This prompted Count Domenico Agusta to offer him a contract, but Hocking refused, saying he wanted more money. The Count was not accustomed to being refused, and this rejection, coupled with Hocking's pledge 'to continue beating you on the MZ', probably convinced the MV boss to give in and offer a new deal. This time it was accepted.

With the likes of Surtees and Ubbiali in the MV team, Hocking did not achieve instant

Gary Hocking winning the prestigious Race of the Year at Mallory Park in September 1961 to round off a magnificent year in which he became 350 and 500cc world champion on four-cylinder MVs.

results; in fact, for some time, he was allowed to race only the 125 and 250cc models, plus occasionally a special 285cc twin. Hocking raced a four-cylinder 350 for the first time at the 1960 Dutch TT, where he finished second to Surtees.

With the retirement of both Surtees and Ubbiali at the end of 1960, there is no doubt that during 1961 Hocking had the field largely to himself. He made the most of the situation, winning both the 350 and 500cc world titles. However, at the final round in Italy, Count Domenico drafted Mike Hailwood into the squad, creating a massive dilemma for Hocking. Having already clinched the world championship, he did not need to win the Italian GP, but, as world champion, it would not help his reputation if Hailwood beat him. The rivalry was intense, for Hailwood had everything to gain and rode hard to win, and Hocking was forced to meet the challenge head on. On this occasion, the Englishman came out on top – after the Rhodesian came off. It was a message to Hocking that the easy days were over and that he was going to have a hard time in 1962 against Hailwood on an identical machine.

The showdown came at the 1962 Isle of Man TT, when Hailwood won the Junior (after Hocking was slowed by a misfire), but Hocking got his revenge in the Senior event (when Hailwood suffered clutch slip). However, the death of his friend, Australian Tom Phillis in the Junior (on a Honda), convinced Hocking that it was time to retire, and he never raced a bike again.

Hocking's tragic death on 21 December 1962 is shrouded in mystery. He was practising for a car race in Natal and the Lotus he was driving simply went straight on when it reached a bend. One source claimed that Hocking died at the wheel from a heart attack. Another report stated that he was suffering from a blood disorder, which caused him to black out and crash. The truth is that no one really knows the cause for sure, and Gary Hocking's death will remain one of racing's mysteries.

(Above) *Gary Hocking (MV) leads Mike Hailwood (Norton) around Gerards Bend during the Race of the Year at Mallory Park, September 1961.*

Hocking's 1961 500 MV four without its fairing. Note fitment of hydraulic steering damper. This and the rear shocks were made for MV by the British Girling company.

(Below) *For 1961, with Surtees and Lightweight champion Carlo Ubbiali retired, only Gary Hocking rode for MV, but the Count chose to run the team under the 'Privat' label, rather than a full works effort.*

MV did not bother to contest the final two rounds at Tampere (Finland) or Buenos Aires (Argentina), but in the 350cc class it was an entirely different matter. Although Hailwood had maximum points after his Isle of Man victory, he did not win another race. Instead, the Honda invasion of the class had begun, with Jim Redman taking the title following four straight wins, in Holland, Ulster, East Germany and Monza. Redman finished at the top of the 1962 350cc championship table, ahead of team-mate Tommy Robb, with MV-mounted Hailwood in third.

Snetterton, 22 April 1962: new signing Mike Hailwood makes his British debut aboard the four cylinder MVs (he is seen here on the 350). In the 500cc event he set a new lap record for the Norfolk circuit at 96.2mph (154.9km/h).

Gary Hocking bump-starts the MV into life at the beginning of the 1962 Senior TT, which he won. It was his last race for the Italian marque.

GILERA COMEBACK

The big news at the beginning of 1963 was the return of Gilera, via a private team run by former hero Geoff Duke. The origins of this comeback can be traced to a meeting the previous year at Oulton Park in memory of the former Gilera rider, Bob McIntyre, who had been fatally injured at the Cheshire cir-

cuit in August that year. As a mark of respect, the Arcore factory had sent one of its 1957 dustbin-faired fours over to the UK for Geoff Duke to do a lap of honour around the circuit. This lap was to spark a Gilera comeback.

In early March 1963, the news was announced that the famous Gilera fours would be back in action, under the private Scuderia Duke banner, with Derek Minter and John Hartle as riders. Their first race would be on 6 April at Silverstone. Even then, there were warning signs that they might not enjoy the fairy-tale ending that almost everyone was expecting. Privateer Phil Read, mounted on a production Manx Norton, split the Gilera duo of Minter and Hartle in the all-important 500cc race.

It transpired that the dreams of the Scuderia Duke were not to come true. First, Minter was badly injured even before the championship season got under way, in an accident at Brands Hatch in early May, riding one of his Norton singles. Read was brought into the team as a replacement. The first Grand Prix, at Hockenheim in West Germany, saw the smaller Gilera four totally outclassed in terms of speed, not just by the full works Honda fours and Bianchi twins, but also by the pre-production Honda CR77 dohc twin. At the second round in the Isle of Man, even though Hartle's circuit experience helped him bring the 350 Gilera home in second place in the Junior TT, the writing was clearly on the wall, and the 350s were soon withdrawn so that the team could concentrate on the 500cc class.

Despite an early-season mauling by Hailwood and the MV, the Scuderia Duke was more competitive in the 500cc class, and the Gileras at last made Hailwood's task more difficult. Even so, their only victory, won by Hartle at the Dutch TT, was made possible only by Hailwood's MV blowing up on the second lap. Meanwhile, Minter had returned to the team in August, but this did not seem to lead to any significant improvement in the Gilera team's fortunes. The scorecard at the end of the season read MV (Hailwood) 7, Gilera (Hartle) 1.

In the 350cc championship, again over eight rounds, Redman retained his title. MV and Hailwood had the satisfaction not only of winning two rounds (East Germany and Finland), but also finishing runner-up.

1964

The first round of the 1964 world championship was scheduled to take place at Daytona Speedway in Florida, USA. Before racing started, MV and Mike Hailwood launched an attempt on the one-hour speed record then held by the late Bob McIntyre (set on a 350cc Gilera at Monza in November 1957). The attempt, made on the morning of 2 February, was an unqualified success, achieving a new one-hour record of 145.675 miles (233.081km) against the Gilera's 142.19 miles (227.519km). That same afternoon, and with only a change of gearing (even the same fairing was used), Hailwood went out and won the 500cc US Grand Prix.

Hailwood's victory was not as easy as many would have predicted. For the first 75 miles (120km) of the race, the undisputed master of the sport found himself embroiled in a tooth-and-nail contest to overcome the little-known Argentinian Benedicto Caldarella. The reigning champion was on his MV, the challenger on a Gilera. Only gearbox trouble finally removed Caldarella from the amazing battle. For lap after lap of the 3.1-mile (5-km) giant speedbowl, the two Italian fours were involved in the closest 500cc battle since the famous Hocking–Hailwood duel of the 1962 TT. It later transpired that the Gilera was one of the Scuderia Duke machines, which had been sent to Argentina at the end of 1963.

The race-following public began to get excited about a repeat performance in the remainder of the championship series, but a series of strikes and other industrial problems effectively ruined any ambitions that Gilera and Caldarella might have had. The result was that the only other clash between these two in the title race came at Monza in September, where the diminutive Argentinian finished runner-up to Hailwood. Caldarella was some ten seconds down, but was

Mike Hailwood (1) keeps his MV in front of Kirby-Matchless-mounted Alan Shepherd, Mallory Park Race of the Year, 1963.

(Below) *For half the race distance in the 1964 US Grand Prix at Daytona in February 1964, the unknown Argentinian Benedicto Caldarella (6, Gilera) challenged the might of Hailwood and MV.*

credited with the fastest lap, at 121.08mph (193.7km/h).

After 1963, Count Domenico Agusta withdrew from the 350cc class, leaving Hailwood with only a 500cc four to race. Even so, Mike-the-Bike still landed the larger-displacement championship, making it a trio of 500cc crowns in successive years.

HAILWOOD'S DEPARTURE

Just as everyone was beginning to believe that MV would shut up shop and quit racing in the face of the rising tide of oriental technology, the wily Count made two important moves. The first was the signing of youngster Giacomo Agostini; the second was the introduction of an all-new 350 triple (*see* Chapter 12), followed soon after by a 500 version.

These moves, although they ultimately led to Hailwood's departure to rivals Honda, put MV back on top, with 'Ago' going on to score a record seven 500cc titles (1966–72 inclusive) and six 350cc titles (1968–73 inclusive) for the Count's team.

Mike Hailwood's final year for MV came in 1965, when he also won his last championship title for the Italian marque – and the last one for the original 500 four – with Agostini finishing runner-up. Hailwood won eight of the ten rounds of the 1965 500cc championship series, including wins in the USA, West Germany, Czechoslovakia, and Italy. Agostini won in Finland (where Hailwood did not finish), while privateer Dick Creith won a rain-soaked Ulster GP on Joe Ryan's Manx Norton.

Hailwood again in record-breaking form, this time at the Dutch TT at Assen on Saturday 30 June 1964. A week of near-heatwave weather ensured another record was broken in front of 200,000 spectators.

Mike-the-Bike Hailwood

Seen by many as the greatest racer of all time, Stanley Michael Bailay Hailwood was born into a wealthy Oxford family on 2 April 1940. His father Stan, a former racing car driver, had made his fortune building up a string of motorcycle dealerships – and playing the stock market. There is no dispute that Stan's cheque book bought Mike early success, fast-tracking his career by months, if not years. However, money can only take someone so far in the racing game, and Mike needed his natural talent to take him the rest of the way.

Mike Hailwood began his racing career at the tender age of 17, his first outing coming at Oulton Park in 1957 aboard a 125 single sohc MV loaned by friend of the family, Bill Webster (who was also a close friend of Count Agusta). During his early career, he rode not only MV singles, but NSU, Norton, AJS, Ducati and FB Mondial machinery. His first works ride was with Ducati (after his father became the British importer!), and his first GP victory, the 1959 125cc Ulster GP, was on one of the Bologna Desmo singles.

In 1961, he not only became a Honda works rider, winning the 250cc world title in the process, but set a new record by winning three of the four solo TTs. Even in the Junior (350cc), he only missed out on winning a fourth when his 7R AJS crankpin snapped.

Hailwood first joined MV officially towards the end of 1961, for the Grand Prix of Nations at Monza in September, where he won the 500cc race. Following the retirement of the 1961 double (350/500cc) MV world champion Gary Hocking after the 1962 TT, Hailwood became MV's star rider until the era of Giacomo Agostini in the mid-1960s. Then, the epic Hailwood (Honda) and Agostini (MV) battles became legendary.

The last time Hailwood climbed aboard an MV was for practice at Monza in 1968; however, on finding he was to play second fiddle to 'Ago', Hailwood quickly switched to a Benelli for the race. By now he was mainly involved in car racing, and this occupation continued well into the 1970s.

In 1978, Hailwood made an historic return to the Isle of Man and had a famous victory on a V-twin Ducati, followed in 1979 by another TT win on a Suzuki RG500. He then retired from racing on either two or four wheels.

The world – and not just the world of motorsport – was shocked when, on 22 March 1981, Mike Hailwood, along with his little daughter Michelle, died in a car crash on their way to collect a fish and chip supper. It was a tragic end to the career of such a great champion.

Giacomo Agostini

Born on 16 June 1942 at Lovere in the Bergamo region of northern Italy, Giacomo Agostini ('Mino' to his close friends, and 'Ago' to his thousands of adoring fans worldwide) got his first taste of motor-cycling in his home village when, aged nine, he would cadge lifts on the petrol tank of a 500 Moto Guzzi single owned by the local baker.

He pestered his father so much that, two years later, Papa finally relented and bought a 50cc Bianchi two-stroke. By the age of fourteen, Ago was racing off road on a 125cc Parilla motocrosser. Four years later, his apprenticeship was complete when he made his competition debut on the tar-mac, on a 175cc Morini ohv single. He finished second in the Trento-Bondone Hill Climb event to the existing Italian champion.

By the beginning of 1963, Agostini was winning races all over Italy on a Morini 175 Settebello. This led factory boss Alfonso Morini to acquire his signature on an official works contract, making him back-up rider to the legendary Tarquinio Provini (formerly with FB Mondial and MV Agusta).

At the end of 1963, Provini quit Morini and signed for rivals Benelli. Ago was promoted to number one. Even before Provini's departure, Ago, now aged 21, had proved at the Italian GP at Monza in September 1963 that he had real class. He had lapped joint fastest in practice with Honda-mounted World Champion Jim Redman and MZ star Alan Shepherd. In the race, the Morini youngster actu-ally led in the early stages, before finally being overhauled and passed by both Provini and Redman. The following year, 1964, Agostini won the Italian Senior Championships on the Morini.

Count Domenico Agusta had always wanted a rider who would be able to win the 500cc world title, but he particularly wanted him to be an Italian. His belief that Giacomo Agostini was the right man led the Count to sign the gifted youngster in time for the 1965 season.

Agostini's first time out did not prove to be the dream outing that everyone had expected. In fact, it was just the reverse, and he crashed out, but all this was soon forgotten. His first win came early in the 1965 season, when he took the flag ahead of the field on one of the new 350 3-cylinder mod-els at the Nürburgring, home of the West German Grand Prix. Agostini then went on to win a record fifteen world titles, thirteen on MVs, the other two on Japanese Yamaha machines. Added to his total of 122 GP victories, it seems like a record that will never be beaten.

Agostini finally hung up his leathers in 1976, at the age of 35. But, of course, the ever-popular Italian superstar did not simply retire and vanish from the scene; during the late 1980s, he became manager of first the Marlboro Yamaha squad, and later of Cagiva.

At the end of 1964, Count Domenico Agusta signed the promising young Morini rider Giacomo Agostini. 'Ago' soon showed his worth by not only leading the 350cc World Championship series, but also putting in a number of appearances in the 500cc category. Here he is shown finishing runner-up to Hailwood at the 1965 Dutch TT on the 500 MV four.

Hailwood's decision to quite MV at the end of 1965 was due to a combination of money and, more importantly, the chance to compete in three capacity classes. Another reason was that Count Domenico had always dreamed of an Italian winning the 500cc title on one of his machines. (For the full story of how this was achieved, *see* Chapter 12.)

As far as the 'old warrior' four-cylinder model was concerned, Mike Hailwood's 1965 championship was really the end of the road. The bike was finally retired at season's end in favour of the lighter three-cylinder mount. The world would have to wait for a new decade – the 1970s – for the emergence of a new breed of four-cylinder MV GP racer.

Count Domenico Agusta (left) with Agostini at the Italian Grand Prix, 3 September 1967. In 'Ago', the Count had finally found an Italian rider capable of winning the 500cc title.

Protar MV Agusta 500cc 4-Cylinder Racing Motorcycle, ⅛th Scale

I am sure that this model will need no introduction to readers of this book; it is one of the world's most famous racing motorcycles, and was ridden by legendary riders such as John Surtees, Gary Hocking, Mike Hailwood and of course Giacomo Agostini. In 1966 another ex-world champion, Tarquinio Provini, had a dreadful accident at the Isle of Man TT (whilst racing a Benelli four), breaking his back. Doctors told him that he would never walk again, though he managed to prove them wrong. He decided to start a factory producing scale model kits of the machines ridden by himself and his adversaries, and this kit has been in the range since the earliest days. As you can see  the detail is excellent, as is the quality of the parts. It all fits together with the minimum of fuss and even a novice would have no difficulty in putting it together. The fairing comes with the aluminium pre-sprayed onto it, making it a simple task to get the colours right. The tyres are made of a rubbery substance, as are the handlebar grips, and all of the electrical and control cables are included in the kit. It is made in ⅛th scale and the finished model is just over 8in (203mm) long and 5½in (140mm) tall. Looking at this model brings back fond memories of the 'Golden Era' of motorcycle road racing.

Ian Welsh

5 Racing Swansong

DEVELOPMENTS AND CHANGES

Towards the end of the 1960s, Giacomo Agostini and MV began a 350/500cc double winning streak. Other, less experienced racing teams might have become complacent, but MV knew well enough that a new challenger could appear at any moment.

At the time, the MV racers were of the three-cylinder across-the-frame configuration, but it was recognized that their life span was limited because of the technical advances being made, most notably in Japan. Work therefore began on the design and construction of a new six-cylinder (*see* Chapter 12). This engine followed traditional MV practice, being an across-the-frame layout, 348.8cc (43.3 × 39.5mm), with the cylinders inclined forward by some 10 degrees from vertical. It had double overhead cams driven by a train of gears, while the primary drive followed the same route as the 1957 500cc six – from one end of the crankshaft.

All the time and effort expended on this project came to nothing when the FIM announced its intention of limiting 350 and 500cc class machines to a maximum of four cylinders and six gears. Besides its six cylinders, the new MV 350 had seven gears. Count Domenico Agusta decreed, therefore, that the whole six-cylinder project should be cancelled in favour of new four-cylinder, six-speed dohc across-the-frame four. However, this was to be done using technology that was more modern than that used on the fours raced during the 1950s and early 1960s, which were by now obsolete.

At the beginning of 1971, significant changes were forced upon the company. In early February, the Count died from a massive heart attack while in Milan. The situation would never be the same again. With his hands-on approach to running the Agusta empire, Count Domenico had been called autocratic, and described as a despot, or even a tyrant. He may have been all these things, and would hardly have been popular with 'modern' management, or trade unions, but he gained achievements where few others would have been able to. He was a man of vision, drive and influence, the like of which has rarely been seen before, or since, in the motorcycle world.

THE NEW FOURS

The first of the new line of fours to arrive was a 350 version, which made its debut at the Italian Grand Prix at Monza in September 1971, ridden by Nello Pagani's son, Alberto. Alberto Pagani already had over a decade of racing experience at the top level, having ridden a variety of machines, including works Aermacchi singles and the twin-cylinder Linto.

On the same day as the new four made its debut, Agostini was forced to retire with engine problems on both the 350 and 500cc

*MV works machinery in 1972.
Note the use of both wire and
cast-alloy wheels. This was the
year when the Cascina Costa
factory responded to the
growing two-stroke menace
with a new 350 four
(349.8cc/53 × 38.2mm), which
had debuted the previous year.
During 1971, Ago had found it
increasingly difficult to beat
the lighter twin-cylinder two-
strokes, but the advent of the
new four gave him the 350cc
world crown in 1972 and 1973.*

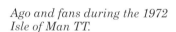

*Ago and fans during the 1972
Isle of Man TT.*

races on the long serving triples, something
which had never before happened in both
classes at the same GP.

In 1972, 'Ago' used both the old three and
the new 349.8cc (53 × 38.2mm) four to gain
yet another 350cc world title, as well as
winning the 500cc title. These titles brought
his total to no less than twelve championships – a figure unheard of at the time for
a single rider.

The new four's other details were as follows: dohc, with aluminium cylinder and heads, sixteen valves, electronic ignition, four 28mm Dell'Orto carbs, wet sump, dry multiplate clutch, six-speed gearbox, gear primary drive, chain final drive, duplex frame with detachable bottom rails, 38mm Ceriani GP front fork, twin shock rear suspension, 18in wheels, triple disc brakes (twin 250mm front discs, single 230mm rear disc), 18-litre fuel tank and four-into-four exhaust system.

However, the 1972 season had not been easy for the Italian pairing of Ago and MV, at least not in the 350cc class. Agostini began the season with the old triple, but after being comprehensively beaten in the first two rounds (at the Nürburgring in Germany and Clermont-Ferrand, France), a switch was made to the four, which was as yet largely untried. As it turned out, the 'Flying Finn' Jarno Saarinen, and his works Yamaha two-stroke twin, posed the biggest threat since the Hailwood-Honda days. The former Yamaha works team captain Phil Read was signed to bolster MV's chances of success; at first he rode only in the 350cc class, and Ago was supported by Alberto Pagani in the 500cc division.

There were twelve rounds in the 1972 350cc championship. Agostini won half of these, with victories in Austria, Italy, Isle of Man, Holland, Sweden and Finland. Jarno Saarinen won three (West Germany, France and Czechoslovakia) with Janos Drapel, Bruno Kneubuhler and Read one apiece.

The final 1972 350cc championship table was as follows:

1st   Giacomo Agostini (MV)
2nd   Jarno Saarinen (Yamaha)
3rd   Renzo Pasolini (Aermachi-Harley Davidson)
4th   Dieter Braun (Yamaha)
5th   Phil Read (MV)

In contrast, the 1972 500cc title race was more of a foregone conclusion, with Agostini and his team-mate Alberto Pagani taking the first two places in ten of the thirteen rounds – although this was done at speeds lower than those in the 250cc class! However, this was to be the last easy year in the 'blue riband' division, as the threat from the Yamaha, Suzuki and the German König two-strokes was about to become very real.

Giacomo Agostini at Brands Hatch on the new four-cylinder MV in October 1972.

Rider's-eye view: the cockpit of a works four-cylinder MV, c. 1973. Note the large white-faced Veglia tacho, which registers 4–18,000rpm, as well as the adjustable damping for front forks, massive steering damper control unit and quick-action fuel-filler cap.

(Below) Agostini on his way to victory in the 1973 500cc Czech GP at Brno.

THE 1973 SEASON

In 1973, although Agostini was to retain his 350cc crown (with victories in France, Italy, Holland and Finland), he found the going much more difficult, both from outside and inside. Team-mate Phil Read stormed home with the 500cc world title.

The 1973 500cc championship comprised a ten-round series. Read won four (West Germany, Holland, Sweden and Spain), Agostini three (Belgium, Czechoslovakia and Finland) and Kim Newcombe (König) won in Yugoslavia. The other two rounds (France and Austria) went to Jarno Saarinen.

The Italian GP at Monza was cancelled after the terrible accident in the 250cc race, which cost the lives of both Saarinen and Renzo Pasolini. There is little doubt that, had he survived, the Finnish Yamaha rider would have posed a very serious threat to MV's chances.

The 1973 500cc championship table ended as follows:

1st   Phil Read (MV)
2nd Kim Newcombe (König)
3rd Giacomo Agostini (MV)
4th Werner Giger (Yamaha)
5th Jack Findlay (Suzuki)

While Agostini had chosen to stay with the existing 500 *tre cilindri* (three-cylinder) bike for the first round in France, Read had been given an experimental 433cc (56 × 44mm) four-cylinder model (based on the 350 four, which debuted in 1971). MV records show that this produced around 90bhp at 12,500rpm, and Read finished the race runner-up to winner Jarno Saarinen's new four-cylinder Yamaha two-stroke.

Read was a non-finisher in the next round in Austria (as was Agostini). In round 3 (at Hockenheim, West Germany), Read debuted and won the race on a larger-engined (498.6cc/58 × 47.2mm) four. However, at some rounds, the old triple was used, either because it was preferred, as in Agostini's case, or because of development glitches, as in the case of Read.

Many asked the question why MV was not at least considering a switch to two-strokes. The answer can be found in the company's long association with the four-stroke. (For much the same reason, Honda developed the vastly expensive oval-piston NR500 in the

On his last MV ride before joining rivals Yamaha, Agostini guns his four-cylinder MV to victory in the 500cc Finnish GP, 29 July 1973.

late 1970s.) Somehow, race-goers would have found it very difficult to accept a two-stroke from MV than one from virtually any other make. Of course, MV had built, sold and even raced two-strokes in the past, but

these were ultra-lightweights. However, as Honda was to find to its cost, a switch was essential if serious championship aspirations were to be fulfilled.

Another factor was the management structure of the company. When Count Domenico Agusta was at the head, it was simply a case of accepting and doing what he wanted. His successors were very different men. Not only did they not really understand the motorcycle racing game, but also, unlike the Count, they were extremely slow at making decisions. No one person was in control, and government officials in far-off Rome had to approve any major spending. Gaining approval was a time-consuming process.

THE 1974 SEASON

After winning the 1973 350cc title, Giacomo Agostini held thirteen world titles and 108 Grand Prix victories, in Grand Prix, a record. A fall at Misano, during a private practice session, prevented him from taking part in the final round at Jarama, Spain. Then came

(Above) *With Agostini's departure, Englishman Phil Read assumed the mantle of MV's number one rider. Seen here with his late wife Madeline in 1973.*

Phil Read (1) sits on the start line at Brands Hatch in late 1973, alongside John Player Norton rider Dave Croxford.

An MV mechanic warms up the 500 MV four at a Brands Hatch meeting in 1974. By now, a glorious era was coming to a close. Compare the paddock scene to today's – no motorhomes, for a start – and note the Jensen Interceptor.

At the 1974 Belgian Grand Prix, Phil Read wheeled out his 'winged' model. Although used in practice, this was not raced, but Read still won for MV for the seventeenth time in seventeen years!

a decision to make a switch to Yamaha, the firm that had given him so many problems in 1973. This meant that he would, in 1974, be riding as principal adversary against his former team-mate Phil Read, who would be racing the marque on which Agostini had achieved all his World Championship titles.

To back up Read, MV signed Gianfranco Bonera. Read's machine was the full-sized 500 four, which was by now fully developed. As on its smaller brother, the cylinders were inclined 10 degrees from the vertical, with square finning, while the engine dimensions were oversquare. The gear-driven double overhead cams, with four valves per cylinder, were fed by a quartet of 32mm Dell'Orto carburettors. Unlike the 350 four, on the 500 a magneto was fitted at the front of the engine and driven by a shaft and gears. The wet-sump lubrication system included a cooling radiator mounted at the front behind the number holder on the nose of the fairing, which was perforated to allow the passage of cooling air.

For 1974, Read and MV were contracted to the French ELF fuel company.

Phil Read MV (1) takes the outside line on his way to victory in the 1974 500cc Finnish GP at Imatra. The other rider is Barry Sheene (Suzuki).

The arrival of slick (untreaded) tyres was a significant development for 1974. Another was the testing of a 'winged' aerofoil device at Spa Francorchamps during practice, although Read chose not to use this fitment in the race. (MV won for the seventeenth time in seventeen years.) In a further development, MV built a larger 580cc engine, for use in non-Championship Unlimited class events, such as on the British short circuits.

There is no doubt that Phil Read (ably backed up by Bonera) had to work hard all season long to ensure that the championship returned to Cascina Costa. This also meant that MV did not contest the 350cc class at all. In effect, ex-teamster Agostini had a clear run at the title, which he duly won.

Ago was less fortunate in the bigger class, and a combination of ill fortune and superb riding by the MV men meant that he finished no higher than fourth in the championship table. Besides Agostini, other major threats that year came from Jack Findlay and Barry Sheene (both Suzuki) and Tepi Lansivuori (Yamaha).

The final 1974 500cc championship table was as follows:

1st Phil Read (MV)
2nd Gianfranco Bonera (MV)
3rd Tepi Lansivuori (Yamaha)
4th Giacomo Agostini (Yamaha)
5th Jack Findlay (Suzuki)
6th Barry Sheene (Suzuki)

(Above) *Read (right) and Sheene together at Mallory Park, 15 September 1974. Read was 500cc champion in 1973 and 1974; Sheene was later the holder in 1976 and 1977, on a Suzuki.*

Read rounding the Mallory Park Hairpin on his factory four-cylinder MV, 15 September 1974.

1974 500cc World Championship Positions

		Fra	WDR	Austria	Italy	IOM	NL	Belg	Swe	Fin	Czech *	Total
1	P Read (MV)	15	0	0	10	0	10	15	12	15	15 (92)	**82**
2	G Bonera (MV)	10	0	12	15	0	8	1	8	12	12 (78)	**69**
3	T Lansivuori (Yamaha)	8	0	0	12	0	12	0	15	10	10	**67**
4	G Agostini (Yamaha)	0	0	15	0	0	15	12	0	0	5	**47**
5	J Findlay (Suzuki)	0	0	8	8	0	0	6	0	8	4	**34**
6	B Sheene (Suzuki)	12	0	10	0	0	0	0	0	0	8	**30**
7	D Braun (Yamaha)	0	0	6	0	0	0	10	0	0	6	**22**
7	P Korhonen (Yamaha)	0	0	0	0	0	4	0	10	6	2	**22**
9	B Nelson (Yamaha)	5	0	4	3	4	0	0	5	0	0	**21**
10	J Williams (Yamaha)	4	0	3	0	0	1	4	1	5	0	**18**
10	C Williams (Yamaha)	0	0	0	0	12	6	0	0	0	0	**18**
12	H Kassner (Yamaha)	0	12	0	0	5	0	0	0	0	0	**17**
12	K Auer (Yamaha)	0	0	5	0	0	5	0	6	1	0	**17**
14	E Czihak (Yamaha)	0	15	0	0	0	0	0	0	0	0	**15**
14	P Carpenter (Yamaha)	0	0	0	0	15	0	0	0	0	0	**15**

*10 rounds – Best 6 rounds count

The 1974 version of the 498.6cc (58 × 47.2mm) 500 MV. Note oil cooler in front of steering head.

The Boxer Flat Four

In its desperate search to remain competitive against the ever increasing rise of the multi-cylinder two-strokes, MV (like Honda, with its ill-fated oblong piston technology NR500 project of the late 1970s) strove to find a four-stroke engine design that would give the necessary power output.

One project was the Boxer flat four. This was the work of Ing. Bocchi, formerly of the Ferrari car company. Although it was, in many ways, a bold and innovative design, it was destined never to turn a wheel.

The main focus of attention was the engine, a water-cooled, flat four, four-stroke with dohc

Designed by Ing. Bocchi, the never-raced Boxer five-hundred fore-and-aft four featured a watercooled dohc engine with 4-valves-per-cylinder and electronic ignition.

and four valves per cylinder. The power unit was mounted with the cylinders jutting fore and aft. A six-speed gearbox cluster was mounted below the crankshaft and there were straight-cut helical primary-drive gears. The four round-slide Dell'Orto carburettors with their integral float chambers were mounted in pairs nearly horizontally above the cylinders. There was a complete lack of finning for either the heads, barrel or crankcases. A massive (again unfinned) sump was cast integrally with the crankcase assembly. Ignition was electronic, manufactured by the Varese-based Dansi company. Designed to run at 15,000rpm, the engine's projected horsepower was 110.

The chassis was of the open type, with the engine mounted at three main points. Suspension was taken care of by a pair of hydraulic rear units supported by a square-section swinging arm, with 38mm diameter front forks. The wheels were of magnesium alloy with seven spokes.

Although extensively bench-tested and mounted in a chassis, the flat four became a victim of Agusta's decision to quit GP racing at the end of the 1976 season.

View of the 1974 engine from the offside clearly showing bolted-up twin front frame downtubes, long sump and magneto mounted at front of crankcases between exhaust header pipes.

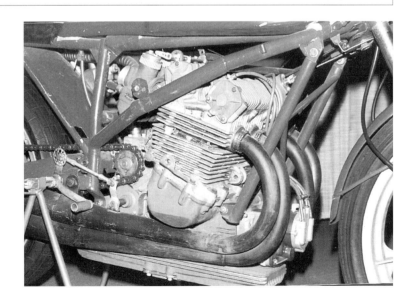

DOWNHILL TO THE END OF THE ROAD

From then on, it was largely downhill all the way for MV, with the two-stroke challenge in full swing, led by Yamaha and Suzuki. Phil Read did his best, but he could only win two (Czechoslovakia and Belgium – the latter for the eighteenth time in eighteen years) of the twelve rounds counting towards the 1975 500cc title.

Technically, MV was almost spent. There was a new single-shock cantilever frame, a host of other chassis experiments, and a prototype liquid-cooled dohc flat four. Many of these experiments, including the flat four, were the work of former Ferrari technician, Ing. Bocchi, but none made any real difference to halt MV's decline.

The 1976 team was sponsored by Italian petroleum giant Api, and run by Giacomo Agostini, instead of the factory.

In the following year, 1976, MV quit competition on an official basis. However, in a very strange set of circumstances, Agostini returned to ride 'privately entered' MVs, backed by the Italian oil company Api. Although the factory was not supposed to be taking a direct interest, there was still a full complement of MV mechanics, including

MV four engine out on the bench, with the clutch removed. Cam boxes and magneto clearly visible.

Arturo Magni

In the MV racing story one man stands out above everyone else – except, of course, Count Domenico Agusta himself. That man is Arturo Magni.

Magni (pronounced 'Maani') rose from a young mechanic to become engineer, then projects co-ordinator and finally team manager. His time with the company spanned its entire four-cylinder GP career, from 1950 until the team's final demise at the end of 1976.

Born in 1920 near Arcore, the home of MV's great rival Gilera, the young Arturo Magni grew up in an Italy that was very much under the boot of Fascist dictator Benito Mussolini. Living under the fascist regime, however, did have one advantage, in that it positively encouraged technical ability, particularly in the field of aviation. In his youth, Magni was an enthusiastic aero modeller, and this interest was to play a vital role in his future, because it led directly to a friendship with Ferruccio Gilera, son of Giuseppe, boss of the famous motorcycle works. When the aircraft factory in which Magni had worked closed its doors in 1947, with the end of the Second World War, and he was made redundant, it was perhaps natural that he should end up in the Gilera racing division. Here, he soon established himself as a mechanic of rare ability. His aviation engineering background proved a great asset, and he was appointed head mechanic in 1948.

In this position, Magni played his part in the success enjoyed by the works Gilera 500 four-cylinder model during the first year of the official FIM-sanctioned World Championship series, in 1949. Gilera's team leader, Nello Pagani, won two of the five rounds and finished the season runner-up to AJS-mounted Les Graham; another Gilera rider, Arcisco Artesiani, was third.

In a strange quirk of fate, Graham, Artesiani and Magni were all destined to join the same new team for the following season, alongside Gilera's chief designer, Ing. Piero Remor. Years later, Magni mentioned in conversation with MV follower Richard Marchant how Remor had told Magni that he was joining MV, and had asked him to accompany him. Remor, Magni and rider Artesiani all came to join MV in the 1949–50 closed season, while Graham was signed by Count Domenico Agusta himself.

In his first couple of years with MV, Magni continued to commute from his Arcore home to Cascina Costa, where the race shop was situated within the main MV factory. He then built his existing home in Samarte, a few miles south of Gallerate. The ground floor is given over to workshops, and an office, which are utilized today for the existing Magni business, with living accommodation above.

In another life, Arturo Magni might well have been a diplomat. One journalist has described him as having 'the ability to parry the most probing – and often loaded – questions with a polite 'yes' or 'no'. And, unlike many of his countrymen, he is both quiet in nature and cool under pressure.' Even today, years after the death of Count Domenico, Magni is guarded in his accounts of the life and times of the team and its owner. However, his stories from those days are much in demand, and he is occasionally persuaded to tell one.

One tale gives a fascinating insight into the racing world of the early 1950s and involves Arturo Magni and the late Bill Webster, a great friend of Count Domenico Agusta, who also raced smaller MVs and later still became the British importer. Webster (known to his many friends as 'Websterini') would collect the MV team from London's Victoria train station and provide them with accommodation. Magni's favourite recollection is of his first visit, when the team arrived at Dover in May 1950 on their way to the TT for MV's Isle of Man debut. Post-war Britain was almost totally devoid of any restaurants that stayed open in the evening, so the entire MV squad was forced to travel in two vans from Dover to Liverpool (some 300 miles), eating only fish and chips. It was a completely new experience for the Italians!

Arturo Magni is also interesting because of the close contact he has had through the years with MV's many star riders. His favourite is Les Graham: 'To me, he stands out above all other riders, and I feel that not enough credit has been given to him. He is the only MV rider whom I have gone on a motorcycle with to a track to actually see the race when it was not strictly necessary for me to be there. And when I say that I had to ride with one hand totally bandaged up after a bad accident in the factory, you can see how much esteem I held him in.'

As for Arturo Magni himself, the following extract from an article in the August 1973 issue of *Motorcyclist Illustrated* magazine, written by David Dixon, best sums up his role within the MV organization, and his wider appeal:

> As a motorcycle team manager and tactician, and above all as a company man, he must surely rank as the motorcycle racing equivalent of the legendary Alfred Neubauer of Mercedes Benz.

500 Four (1976)

Engine

Type	Air-cooled dohc four, across-the-frame
Bore and stroke	58 × 47.2mm
Capacity	498.6cc
Compression ratio	11.2:1
Carburation	Four Dell'Orto 32mm E154 carburettors with integral float chambers
Lubrication	Wet sump
Max power	98bhp @ 14,000rpm
Fuel tank capacity	4 gallons (18 litres)

Transmission

Gearbox	Six speeds
Clutch	Dry, multi-plate
Primary drive	Gear
Final drive	Chain
Ignition	Magneto

Frame

	Tubular steel, duplex cradle

Suspension and steering

	front	Teledraulic fork
	rear	Swinging arm with hydraulic shock absorbers
Tyres	front	3.50 × 18
	rear	4.75 × 18
Brakes	front	Twin 250mm discs
	rear	Single 230mm disc

Dimensions

Dry weight	264.5lb (120kg)

Performance

Top speed	177mph (285km/h)

The *nuovo* four-cylinder MV replaced the long-serving three-cylinder in time for the beginning of the 1973 season. At first it had a 433cc (56 × 44mm) engine size, before being brought up to a full 500. The early examples (1973) also featured wire wheels.

In 1974, a new frame was adopted, with tubes running from the cylinder head to the swinging-arm pivot, in a style pioneered by British racer/manufacturer Colin Seeley. Cast-alloy wheels were fitted, along with a disc in place of a drum on the rear. The swinging arm was also modified.

In 1975, experiments were carried out with a frame featuring multi-adjustable angles for the twin rear shocks. In addition, a Yamaha-type horizontal monoshock frame was also used; an interesting feature of this latter type was that small hydraulic dampers were also employed at each side of the rear wheel.

For 1976, with sponsorship from Marlboro, the frame was again changed, reverting to a traditional duplex cradle affair with detachable bottom rails for easy engine removal. Twin rear shock absorbers were also specified. Besides its new Team Marlboro colour scheme, other changes included the fuel tank and the exhaust, with the megaphones being not only upswept but kinked towards the end farthest away from the exhaust pipes. The tyre combination that year was usually a slick rear and treaded front. Giacomo Agostini took the last-ever MV 500cc Grand Prix with one of these machines, at the Nürburgring in August 1976.

The final 1976 version of the MV 500 four. This featured frame modifications, including a more traditional duplex frame structure, revised fuel tank and upswept meggaphones.

(Right) *Experimental four-valve-per-cylinder head.*

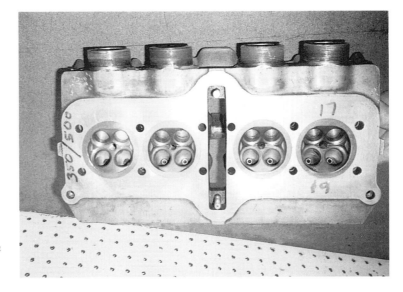

In its final form, the 500 produced 98bhp at 14,000rpm and 177mph (285km/h), but even this speed was not enough to beat the two-stroke opposition regularly.

Arturo Magni, and a new, improved four-cylinder model. There is little doubt that this motorcycle was very rapid, but it was also exceptionally unreliable, at least at the beginning. On the odd occasion when it did last the distance, however, it was a potential race winner, and Agostini took it to victory in the Dutch TT at Assen. MV Agusta's last Grand Prix victory came at the final event in the 1976 series, the 500cc race at the Nürburgring, West Germany.

Apart from 'demonstration' appearances, including one by Agostini at Brands Hatch, and Phil Read at Cadwell Park, this really was the end of the road. The accountants who controlled the Gruppo Agusta purse strings had the final say, leaving enthusiasts the world over with only their memories. Those fabulous race machines, which had given MV a record number of championship and race victories, were placed under lock and key at the Cascina Costa works. Fortunately, some of these machines have found their way into private collections and can be enjoyed by today's generation of racing enthusiasts at Classic events.

The last MV Grand Prix victory came at the final round of the 1976 season – the 500cc race at Germany's Nürburgring. Giacomo Agostini kept his four ahead of the two-stroke pack to win.

MV's record remains unmatched until the present day: 75 world championship titles (riders and manufacturers), 270 Grand Prix victories, and no less than 3,027 international race wins. On present form Honda are the ones most likely to eventually beat it.

MV World Champions

1952	125cc	Cecil Sandford	1969	350cc	Giacomo Agostini
1955	125cc	Carlo Ubbiali		500cc	Giacomo Agostini
1956	125cc	Carlo Ubbiali	1970	350cc	Giacomo Agostini
	250cc	Carlo Ubbiali		500cc	Giacomo Agostini
	500cc	John Surtees	1971	350cc	Giacomo Agostini
1958	125cc	Carlo Ubbiali		500cc	Giacomo Agostini
	250cc	Tarquinio Provini	1972	350cc	Giacomo Agostini
	350cc	John Surtees		500cc	Giacomo Agostini
	500cc	John Surtees	1973	500cc	Phil Read
1959	125cc	Carlo Ubbiali	1974	500cc	Phil Read
	250cc	Carlo Ubbiali			
	350cc	John Surtees			
	500cc	John Surtees			
1960	125cc	Carlo Ubbiali			
	250cc	Carlo Ubbiali			
	350cc	John Surtees			
	500cc	John Surtees			
1961	350cc	Gary Hocking			
	500cc	Gary Hocking			
1962	500cc	Mike Hailwood			
1963	500cc	Mike Hailwood			
1964	500cc	Mike Hailwood			
1965	500cc	Mike Hailwood			
1966	500cc	Giacomo Agostini			
1967	500cc	Giacomo Agostini			
1968	350cc	Giacomo Agostini			

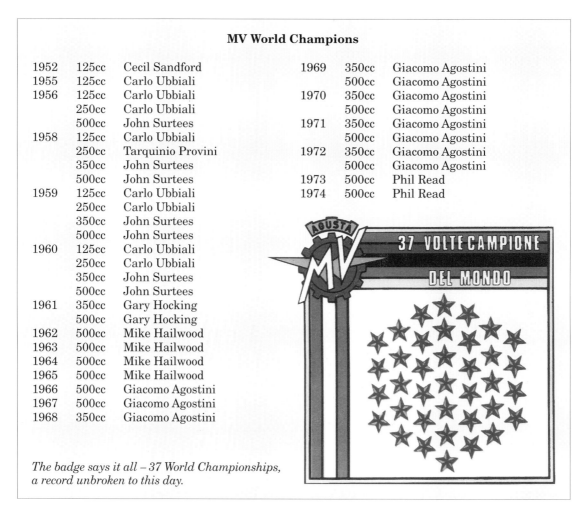

The badge says it all – 37 World Championships, a record unbroken to this day.

Phil Read, MV's Last Champion

To many, Phil Read is a hero; to others, he is just the reverse. The fact is that this Englishman, born in Luton on 1 January 1939, was 'combative and stubborn by nature' (as described in *MV Agusta*, by Mario Colombo and Roberto Patrigani, Giorgio Nada Editore).

Read first leapt to fame by winning the amateur Manx Grand Prix in September 1960 on his privately entered Manx Norton single. After some excellent performances on 350/500cc Nortons on the British short circuits, he was singled out in 1963 by the Scuderia Duke Gilera squad (originally to replace the injured Derek Minter), before being snapped up to ride the RD56 Yamaha two-stroke twin. Again, someone else's misfortune (in this case, Tony Godfrey's 1963 TT crash) opened the door for him, bringing him a full Yamaha works ride.

Read proved that he deserved the breaks, because he went on to become Yamaha's number-one rider with no less than four world titles (three 250cc, and one 125cc) before the Japanese marque finally pulled the plug on its GP effort at the end of 1968. Read rode with Benelli, before gaining another 250cc world title, this time the 1971 championship, aboard a Fath-tuned privately entered TD Yamaha.

At this time, MV was on the eve of launching a brand-new four-cylinder model. In the face of an ever-expanding Japanese challenge (led by Yamaha's Jarno Saarinen), the Gallerate factory decided that it needed more firepower; Angelo Bergamonti had lost his life at Riccione, and Alberto Pagnani was seen as a support rider. Phil Read was signed, in what turned out to be an inspired move. For a start, it had the effect of shaking up Agostini's previously held views that he could take things easy – MV now had two world-class stars, not one.

In 1972, Read took part in only the 350cc class, winning on the MV at the East German GP (Sachsaring) and gaining several other good placings. For 1973, he joined the 500cc division, where he went on to prove his worth by become the new champion at season's end. With Ago third, behind König's tragic Kim Newcome (killed at Silverstone in August), the sparks began to fly. Ultimately, Agostini upped and left, joining rivals Yamaha.

Read repeated his and MV's success again in 1974, and remained for 1975, although in his last year the best he could do was win in Belgium and Czechoslovakia. When these wins were added to the other points he had scored, they placed him second in the 500cc table behind Agostini and Yamaha.

Agostini's final outing on the MV came at Brands Hatch in the 1976 Race of the South meeting.

Giacomo Agostini Scorecard

World Titles

1966	500cc (MV)
1967	500cc (MV)
1968	350 and 500cc (MV)
1969	350 and 500cc (MV)
1970	350 and 500cc (MV)
1971	350 and 500cc (MV)
1972	350 and 500cc (MV)
1973	350cc (MV)
1974	350cc (Yamaha)

Grand Prix Victories

350cc	54 (48 MV, 6 Yamaha)
500cc	68 (62 MV, 6 Yamaha)

Isle of Man TT Victories

350cc	5 (1966, 1968, 1969, 1970 and 1972 – all MV)
500cc	5 (1968, 1969, 1970, 1971 and 1972 – all MV)

First 350cc GP win: 1965 German, Nürburgring (MV)
First 500cc GP win: 1965 Finnish, Imatra (MV)
Last 350cc GP win: 1976 Dutch, Assen (MV)
Last 500cc GP win: 1976 German, Nürburgring (MV)

Riders Contracted by Factory to Race MVs

125cc	250cc	350cc	500cc	
Graham*	Lomas	Graham*	Artesiani	Masetti
Bertoni	Masetti	Lomas	Armstrong	Surtees
Matucci	Taveri	Dale	Leoni	Columbo*
Sandford	Venturi	Franzoni	Graham*	Shepherd T
Copeta	Ubbiali	Masetti	Bandirola	Shepherd A
Sala	Columbo*	Bandirola	Bertacchini	Gonzalez
Lottes	Libanori	Surtees	Amm*	Hartle
Salt	Hartle	Columbo*	Lomas	Venturi
Williams	Chadwick	Shepherd T	Forconi	Cantoni
Ubbiali	Provini	Shepherd A	Giani	Mendogni
Forconi	Hocking	Hartle	Gugielminetti	Hocking
Pagani N	Milani	Brambilla	Francisci	Hailwood
Taveri		Venturi	Muller	Grassetti
Columbo*		Hocking	Milani	Agostini
Atorrasagasti		Hailwood	Sandford	Bergamonti*
Masetti		Agostini	Dale	Read
Venturi		Bergamonti	Pagani	Pagani A
Libanori		Amm*	Taveri	Bonera
Chadwick				
Gonzalez				
Provini				
Vezzalini				
Hocking				
Spaggiari				
Magi*				
Nencioni*				

* Riders who lost their lives while competing on MVs

6 Early Roadster Fours

The First Four

Unlike elsewhere in Europe, Victorian England was still very much a society dominated by horse-power. There was a vast array of draconian regulations impeding the development of self-propelled vehicles, including a speed limit of 4mph (6km/h). It was only in 1896 with the passing of the Emancipation Act that these restrictions were lifted and the new transport industries began to flourish. Despite this, it was an Englishman, Major (later Colonel) Henry Capel Lofft Holden, who built the world's first four-cylinder motorcycle, in 1897.

Variable gears were still well over a decade away, so Holden relied instead on sheer power to overcome the inherent inflexibility of direct drive. Like the German pioneers, Hildebrand and Wolfmüller, he employed long, exposed connecting rods and cranks linked directly to the rear-wheel spindle; however, Holden's engine differed significantly from the Munich-made device.

The four cylinders were positioned in horizontal pairs, each like a straight unpinned pipe, closed at both ends. In each pipe was a long piston with a crown at each end, and explosions took place at alternative ends of the cylinders, giving an impulse on each stroke. With a total capacity 1047cc (54 × 114mm), massive gudgeon pins or 'crosshead' pins, as used on a steam railway engine, projected through slots in the cylinder walls to engage with the connecting rods. Automatic inlet valves were employed, but the mechanically operated exhaust valves, lacking any convenient rotary motion in the engine to drive them, were actuated via a camshaft, chain-driven from the machine's rear wheel.

The four-cylinder Holden engine operated at a mere 420 revolutions per minute, at which it produced 3bhp, giving it a top speed of 24mph (39kmh). Limited production was carried out by the Kennington, London-based organization, but a change to water cooling was required as early prototypes suffered from overheating problems. Originally, production had been planned to commence in 1898, but it was not until the following year that it actually got under way. Unfortunately, water cooling added to the already considerable weight of the Holden four, so it fared badly against the smaller cars, which offered superior comfort, reliability and price. However, these examples were exhibited at the 1901 Stanley Show, and one was ridden from London to Petersfield and back, a total distance of 106 miles (170km), without trouble; but this marked the swansong of the world's first Superbike. An 1897 prototype of the Holden survives today in London's Science Museum.

PUBLICITY AND RUMOURS

Count Domenico Agusta was someone who appreciated the value of publicity, and for many years he led the world's motorcycle press a merry dance. He danced to a tune with which they simply could not keep up – it concerned the elusive road-going version of his company's fabulous Grand Prix racing four-cylinder models.

The 1950 R19

The twenty-eighth Milan Show opened on Saturday, 3 December 1950 in the Palazzo Triennale, its traditional pre-war home. Among the huge tide of exhibits, one entry stood head and shoulders above the rest. This was a prototype 500 four-cylinder road-going MV Agusta, the R19 Gran Turismo. As *The Motor Cycle* said in its 7 December 1950 show report: '[It] is a very handsome job indeed with its enormous centrally disposed brake drums, twin headlamps, long racing seat, twin revmeter and speedometer in the tank, and general air of enormous urge; it should be the Vincent of Italy.'

The Gran Turismo, together with its racing brother, was the work of Ing. Piero Remor, who, together with race mechanic Arturo Magni, had left rivals Gilera some twelve months earlier.

Perhaps it was inevitable, but the MV quattro more closely resembled the Gilera of that era, but probably to avoid being accused of simply transferring technology from one paymaster to another, Ing. Remor had incorporated several new features in the design. These included shaft final drive, two 27mm SS1 Dell'Orto carburettors (in other

The MV Agusta stand was often used to display one-off show bikes alongside the company's production roadsters and factory racers.

words, two cylinders sharing one instrument), and revised suspension. On the 1950 model factory MV four-cylinder racers, there was torsion-bar suspension, both fore and aft. However, on the 1951 model, and the road-going Gran Turismo, telescopic front forks were the order of the day. There was also a parallelogram rear fork (double swinging arm). This last feature was to

The 1950 R19 – a dream bike that remained just a dream; this gran turismo four-cylinder roadster never entered production.

help tame that Achilles heel of shift final-drive torque reaction.

The engine shared the square 54 × 54mm bore and stroke dimensions of that year's racer, giving a displacement of 494.4cc. Other features, also shared with the racer, were the engine's crankcases, cylinder heads, dohc and magneto ignition. In many ways, unlike the eventual 600 roadster launched fifteen years later, the R19 was a streel-legal MV racing four, but given some neat styling touches – such as the twin headlamps and tank-mounted instrumentation.

The R19's engine was clearly derived from the original 1950 500 GP bike, complete with shaft final drive and twin carbs. There was one silencer on each side – the exhaust header pipes being siamezed.

The all-chrome exhaust was a 4-into-2, while the brake hubs and fork sliders were highly polished aluminium; as were the welled Borrani-made 19in wheel rims. As on the racer, the engine castings were of an unpolished matt finish. Brake drum diameters were 230 and 220mm front and rear, respectively.

Did Count Domenico Agusta ever really intend to sell this machine to the public? Even in those austere post-war days, there would have been takers for what would have been an expensive hand-built flagship. In tribute to the R19, it has to be said that, even fifteen years later, it was much more desirable than the ugly 600 of 1965.

As early as the Milan Show at the end of November 1950, the Count was wooing the pressmen with a delectable Gran Turismo MV roadster four, coded the R19. Ultimately, the fate of this attractive machine was to be a show-time publicity piece hawked around the European show circuit. Eventually, even the most enthusiastic supporter had grown tired of promises to bring the roadster into production, and by the end of the 1950s it was virtually forgotten.

Suddenly, in the early 1960s, rumours once again began to circulate that MV Agusta was developing a street version of its victorious four-cylinder GP bike. This time, the stories were largely mistrusted, both by the journalists who wrote articles based on the leaked information, and by their long-suffering readers. The company had cried 'wolf' just once too often, so, when a real-live prototype of a brand-new roadster four made its bow a couple of days prior to the start of the bi-annual Milan Show, in late 1965, many people were caught napping. The crafty Count had once again stolen a lead and pulled off a massive PR coup.

DESIGN

However, those enthusiasts who had hoped for a road-going replica of the MV racers that had been ridden to victory by the likes of Surtees, Hocking and Hailwood were to be sadly disappointed. Instead, what they got was one of the world's ugliest-ever motorcycles, even though there was no doubt that with such an impressive engine the machine had presence.

Undoubtedly, in order to prevent racers from creating their own four-cylinder MV, the Count imposed certain restrictions on the design: the engine displacement was almost 600cc; the engine tune was low; the bike had touring cycle parts; and the final drive was by shaft. Even the colour scheme

The prototype 600 four-cylinder roadster as displayed at the 1965 Milan Show must surely rank alongside the most ugly motorcycles ever made; this is the production version at Milan two years later.

was as far removed as possible from the fire-engine red of the racers – the roadster was a sombre black.

From a styling point of view, the road-going bike did not even look Italian. Instead, with its mass of humps and angles, together with lashings of garish chrome, the 600 MV four looked more like some of the worst of the Japanese offerings of the day. Certain fittings also appeared that were likely to have been fitted for the US market, such as the 'cowhorn' handlebars, a stepped dual saddle and crashbars as standard equipment.

PRESS COMMENTS

By the time the roadster was actually put on to the market, the styling had been tinkered with to make it a slightly more attractive machine. Even so, the American *Cycle World* magazine had this to say in a 1968 issue:

> The appearance of the 600cc MV Agusta is something we'll leave to your own taste. We feel it is far too dated for the modern MV concept but at least it's different, ranging from that giant rectangular headlight to

the hump-backed gas tank or the stair-stepped seat.

Cycle World was to conclude that the ultimate tragedy of the 600 was that, behind its idiosyncratic styling, it was technically a real pace-setter. It was the first of the modern across-the-frame fours; one of the first machines to use electric starting; one of the first to experiment with disc brakes; and one of the very few for many years to come to offer the luxury and convenience of shaft final drive. Furthermore, whatever anyone's opinion of its weird looks, the beauty of its engine assembly was never in doubt. Instantly recognizable as associated with its GP brothers, it was in many ways almost worth putting up with the look of the motorcycle just to have that legendary engine in your garage.

In many ways, this was the attitude adopted by the popular press, who choose largely to ignore the newcomer's appearance (and its performance limitations), and instead focused their journalistic attention on the engine's racing heritage, the brand name on the fuel tank and the thrill of actually riding a four-cylinder MV on the highway. One typical headline described it as 'the most fabulous roadster ever produced'.

The 1965 Milan Show

By the mid-1960s, the Italian motorcycle industry, faced with a falling home market, was being forced for the first time to place greater emphasis on exports. This became evident at the thirty-ninth International Milan Show, which opened on 4 December 1965.

For the first time in many years, big bikes were very much in evidence at Milan in 1965. The British industry was well represented, with an attractive composite stand on which AJS, BSA, Norton Matchless and Triumph displayed a good selection of big British models, together with dirt bikes from the likes of Greeves and BSA. The Italians had some big machinery of their own on show. Ducati displayed the massive Berliner-inspired 1260 Apollo V-four (which was destined never to enter production), while on the Moto Guzzi stand there was the 700cc V7, the first of the now famous V-twin line with shaft final drive and electric start. A 500 Triumph-engined machine named the Grifo, by the Bologna-based Italjet concern, really caught the eye. This featured a duplex loop frame, Ceriani suspension and Grimeca full-width drum brakes.

However, MV really had the last word. For the first half of the show, they displayed two empty platforms 'reserved for surprises'.

The front page of *Motor Cycle News* dated 15 December 1965, said it all: 'MV Spring a Roadster Four; Milan Show Sensation'. The article went on to say, 'A glittering piece of MV showmanship, in the shape of a 600cc four-cylinder shaft-driven roadburner, stole the limelight at the Milan Show, in Italy, last Wednesday.'

The other Agusta mid-show entry was the 250 Bicilindrica, a pushrod twin-cylinder machine with five-speed gearbox. Unlike the 600 four, of which a mere 135 were subsequently built, the smaller machine was constructed in relatively large numbers, together with a later 350 version.

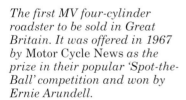

The first MV four-cylinder roadster to be sold in Great Britain. It was offered in 1967 by Motor Cycle News *as the prize in their popular 'Spot-the-Ball' competition and won by Ernie Arundell.*

PRODUCTION BEGINS

At the time of the 600's Milan launch, the biggest question was, 'Will it really ever reach production?' Those pundits who doubted (and they had every right to, based on past performance) were soon to have their answer. The first production machines, in a batch of twenty, went on sale in the summer of 1967 at a cost of 1,160,000 lire each.

These production bikes displayed several minor changes in specification from the 1965 Milan Show prototype. For a start, the carburettor type was switched from Dell'Orto SS1 to UBF; the pair of silencers at the end of the siamezed exhaust were not only of a totally different shape, but also much longer; the battery box was larger; the wheels now carried welled Borrani alloy rims, instead of round-section chrome-plated ones; and the mudguards were not only of a different shape, but the rear one was also considerably longer. Both brakes had been modified, with rod instead of cable operation at the rear, while the calipers for the twin front discs were now mounted behind the fork legs. Modifications were also made even to the external appearance of the engine, even if these were minor rather than major.

None of the above made much difference to the fact that the 600 MV four was a tourer, and an expensive, ugly one to boot. Even so, when MV released test examples, there was a stampede by journalists around the world (not matched by one on the part of the really important people, the paying customers), intent on satisfying their own ego, and on being the first to publish a test of a road-going four-cylinder MV.

THE ENGINE

The basic dohc four-cylinder engine was without doubt clearly based on, if not identical to, the world-beating Grand Prix winners, but it was certainly no performance unit. The mood of the era is captured by the following extract from the April 1968 edition of *Cycle*: 'One must conclude that Count Agusta wanted to produce an untamperable engine, but just couldn't help making one that twangs the mainstrings of a tuner's imagination.'

Close-up of the production 600 four as displayed on MV's stand at the Turin Show, April 1967.

MV's design team's initial brief in meeting the Count's ambition for his roadster had been to up the engine displacement to 591.8cc. This had been achieved by employing a bore and stroke of 58 × 56mm, both of which were different from the various MV Agusta 500 Grand Prix racers. A quartet of four separate aluminium cylinder barrels were specified and these were inclined forward at an angle of 20 degrees. They were equipped with austenitic steel liners, although strangely these pressed-in sleeves were not to be made available as original factory spare parts. The pistons were of a high-domed, forged, three-ring type, and, with such a high compression ratio of 9.3:1, premium-grade fuel was needed.

The cylinder head was a one-piece casting, with the centre serving to locate the top of the timing tunnel, which ran up between cylinders two and three to provide the drive to the twin overhead camshafts, which each ran in four bearings. This tunnel contained a set of three matched, straight-cut timing gears, supported by a cast-iron gear carrier, and driven off the centre of the crankshaft.

As with the vast majority of MV Agusta engines, the light-alloy head featured cast-iron valve guides and valve seats, with the valves themselves inclined at 40 degrees and each featuring twin coil springs. The inlet head was a large 30mm in diameter while the exhaust was smaller, at 28.6mm. Camshaft lift was mild at 8mm (the same as later used for the early 750S model) and on the 1965 prototype, the valve clearances were adjusted with shims sitting on top of a bracket, which fitted over the valve stem. In fact, only the first batch of fifty production examples were to employ this type of valve gear, while the remainder of the 600cc engines that followed featured shims that were more securely (but more awkwardly) positioned under the bucket and on top of the valve stem.

At the base of the engine, the crankcase followed traditional MV four-cylinder practice, which was rather unconventional to the untrained eye. The crankcase itself, for example, was a one-piece casting almost up to the base of the cylinders and included a massively finned 3-litre sump. Across its top face was a separate, cast-alloy crankshaft carrier, from which the crank hung on two ball bearings, plus six split-bearing mountings that projected down into the case. This design meant that there was no risk of leakage along a central crankcase joint, and also meant that any mechanical work on the bottom end of the engine would be much easier. This crank carrier was bolted together with twelve 7mm studs, and undoing these allowed the crankshaft, cylinder barrels, heads and cam drive to be removed completely.

The crank itself was a composite (built-up) type, supported in the carrier by four inner and two side main bearings. This crankshaft design, with its separate crankpins, was exceptionally robust and allowed the employment of traditionally Italian one-piece connecting rods with caged roller big-end bearings, while 17mm gudgeon pins ran directly into the con-rod eye itself.

With crankpins set at 180 degrees, and ignition occurring at 180-degree intervals, the firing order in the cylinders was 1-3-4-2. The ignition was provided with a battery (12-volt, 18 amp/hour) and coil, fired by a German

Nearside view of one of the early 600 four production batch.

Bosch distributor with a steel body, which contained a single set of contact breakers and condenser. This assembly was mounted vertically behind the cylinder barrels, and nestled between the carbs, which on the production version were a pair of Dell'Orto UBF 24BS instruments, compared to the prototype's SS1 27s (borrowed from the racer).

The race-derived motor could spin to over 9,000rpm, but, in keeping with its low state of tune, maximum power was limited to 52bhp at only 8,200rpm. During an American test conducted by *Cycle World* magazine, the maximum speed attained was a shade below the magic ton – 99.3mph (158.88km/h). Perhaps having only two tiny 24mm carburettors, which might just have been suitable for a two-fifty twin, certainly did not help the performance of a 600 four. The lack of thought extended to the fact that no form of air filtration was provided for what was supposed to be a touring, rather than a sports bike.

The turbine smoothness of the engine and the superb sound from the exhaust went some way to offsetting the deficiencies. Another bonus for its role as a tourer was the shaft drive; in those far-off days, very few motorcycles had this feature.

THE TRANSMISSION SYSTEM

The transmission system comprised a wet multi-plate clutch mounted on the near side and driven by helical cut primary gears with a ratio of 1.75:1. There was a total of 12 clutch plates: six drive and six driven. These were retained by five screws, which had to be precisely adjusted to ensure smooth clutch operation. All five screws had to be the same correct length, and all of them had to be done up fully to ensure that clutch operation was not impaired. In practice, gear shifting on the MV could scarcely be faulted – something of a rarity for shaft-driven motorcycles of the period.

The five-speed gearbox had ratios of 3.57, 2.43, 1.68, 1.24 and 1:1, with an additional reduction in the shaft drive. This was turned through 90 degrees in the gearbox casing by bevel gears with a ratio of 1.066:1, while the rear hub bevel ratio was 1.33:1. As on the contemporary BMW flat-twin range, the shaft itself ran in the offside leg of the steel swinging arm.

Electrics

At the base of the engine was a 12-volt, 135-watt dynamo and the electric starter, with the starter positioned to the rear underneath the swinging-arm spindle. The drive to both was by one-way rubber belts, with an inner one that drove the generator, while the outer was for the starter. Inside the engine itself, the starter shaft doubled up to drive the oil pump, which was of the gear type and similar to that of the factory racers, but mounted horizontally instead of vertically. The oil pump drive gear on this shaft was only a press-fit, with no splines or keys. This says something about MV's engineering tolerances.

There is no doubt that, for its chosen task, the 600cc MV engine was generally over-engined and could therefore achieve high mileages without showing signs of distress, except for one notable exception. This concerned the transmission of power via the gearbox to the shaft final drive. The gearbox was, in fact, modified from the 101 production model onwards, since originally the first and second gears on the mainshaft were separate components. Afterwards, they were manufactured integral with the mainshaft. Even so, this modification was not without its problems, and both methods were to suffer from the same weakness – that if too great a load was placed on the drive, the weak link would expose itself. This centred on the needle roller-bearing

Silencers on the production 600 four differed from those on the original 1965 prototype.

DISC BRAKES

The twin mechanically operated Campagnolo disc brakes on the front wheel represented a definite weakness that did afflict the 600 four. Unlike later hydraulic systems, they were both crude and inefficient. According to one magazine, 'the ones fitted to the 600 MV would hardly stop a racing pedal cycle, let alone suffer the stress of pulling up an over-500lb mass from 100mph'. MV considered it worth fitting these brakes at a time when discs were in their infancy, but the company did eventually see the error of its ways. For the first batches made from early 1971 onwards, the Campagnolos were ditched in favour of a massive Grimeca drum brake.

Luckily, the rear drum, a single leading shoe full-width affair was both powerful and reliable – and all the better for its conversion from cable to rod operation by the time the 600 entered production.

EXPORTS TO BRITAIN AND THE USA

The 600 four was certainly comfortable, provided, of course, that the owner was satisfied to ride around at a leisurely pace. Unfortunately, the lack of low-down torque and the hi-revving nature of the engine rather conspired against taking things easy. This shortcoming was highlighted by the experience of the first owner of an MV 600 four to be imported into Britain. Ernie Arundell, who had been lucky enough to have won the machine via a *Motor Cycle News* competition, was a sidecar enthusiast, and equipped his prize with a third wheel. This did not work, and the MV and chair went up for sale only a few months later.

The first 600 MV to reach North America was imported into the USA by Californian businessman Gilberto Cornacchia. After it

supporting fifth gear at the bevel drive pinion in the gearbox cover. The bearing mounted in the nose of the mainshaft suffered from a lack of lubrication and, under heavy usage (fast road work, or racing), could be prone to collapse.

In its intended touring role, the 600 did not suffer from this problem at all, but any attempt at tuning, and when the same layout was used on the later sportsters, the transmission problem came to be regarded as the Achilles heel of all the road-going production fours. It was this same weakness that put paid to Agostini's hopes of success in the 1972 Imola 200 aboard a race-kitted shaft-drive 750 MV Formula 750 bike, and led to his ultimate retirement.

600 Four (1967)

Engine

Type	Air-cooled dohc across-the-frame four-cylinder, with gear-driven cams
Bore and stroke	58 × 56mm
Capacity	591.5cc
Compression ratio	9.3:1
Carburation	Four Dell'Orto MB24 carbs
Lubrication	Wet sump, gear pump
Max. power (at crank)	52bhp @ 8,200rpm
Fuel tank capacity	4.4 gallons (20 litres)

Transmission

Gearbox	Five speeds
Primary drive	Gear
Final drive	Shaft
Ignition	Battery/coil, 12 volt

Frame

Tubular steel, full duplex

Suspension and steering

Suspension	front	35mm, fully enclosed, teledraulic
	rear	Twin shock, with enclosed springs
Wheels		18in wire with Borrani welled alloy rims
Tyres	front	3.50 × 18
	rear	4.00 × 18

Brakes

	front	Twin mechanically operated 214mm Campagnolo discs
	rear	SLS 200mm drum

Dimensions

Length	84.5in (2,110mm)
Wheelbase	55.5in (1,390mm)
Ground clearance	6in (150mm)
Dry weight	486lb (221kg)

Performance

Top speed	106mph (170km/h)

had been put through its paces for the March 1968 edition of *Cycle World*, it was offered for sale in the classified section of the same magazine for $3,500. Meanwhile, Ed La Belle (one of the early road-racing pioneers, who campaigned during the 1950s a BMW flat-twin, among other machines, in the USA, Canada and the Isle of Man), was appointed MV's official importer for the North American continent. In 1968, his company, Ed La Belle Cycle Engineering of Philadelphia, listed the 600 MV at $2,889.

SPECIALS

Although Count Domenico Agusta had done much to try to prevent customers tampering

The final batches of the 600 four built in 1971 and 1972 featured a double-sided 4LS Grimeca-made drum brake in place of the less than satisfactory Campagnolo mechanically operated twin-disc set-up.

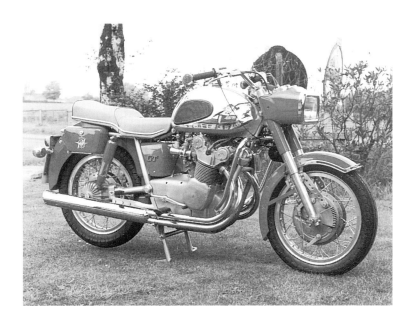

with and even racing the 600, several people did just that. Besides a Swiss sidecar (*see* Chapter 11), two other notable efforts were made. The first came in 1969, when the Swedish international-class racer Sven Gunnarsson produced a very neat, but ultimately unsuccessful, solo racer. This was converted to 500cc, given chain final drive, a quartet of 26mm Amal concentric Mark 1 carbs, a one-off frame, a set of Rickman Metisse front forks and a disc front brake from the same source. The rear brake was an Italian 200mm Fontana drum.

Probably the most successful of the MV 600 specials were those built by Massimo Tamburini at the beginning of the 1970s, before he started the Bimota concern. In fact, it was his success at building a creditable sports bike out of the touring 600 that first brought him to the attention of the Italian motorcycle press.

Observing that people would go to such extraordinary lengths to create a sportster probably led MV to take that route itself.

A SALES FLOP

The touring route also proved not to be a commercially viable way of marketing the four-cylinder MV. During its entire production life, which spanned almost seven years, only a paltry 135 examples were sold. Of these, less than thirty-five had been shipped across the Atlantic to the USA.

Before his death in 1971, the Count must have read the press reports, and recognized the error of his original decision of offering such an ugly-duckling rather than a glamorous sportster. However, it often seemed as if the journalists were trying not to hurt his feelings, as the following comment in *Cycle* in 1968 suggests: 'By building a version of his fabulous dohc racing four engine into a touring frame, Count Agusta has created a unique, weird and wonderful collector's dream.' Surely the truth was really that it was the *wrong* bike? But the wrongs were righted, and the machine that should have been built duly arrived at the beginning of the 1970s, in the shape of the sexy and stunning 750S.

7 750S

STYLE AND STREET CRED

Critics (and there were many) of the 600 four-cylinder roadster were largely silenced when Agusta staff wheeled out the new 750S at the Milan Show in November 1969. Here at last was an MV four street bike that could hold its head up with pride. The 600 was staid, unexciting and ugly; by contrast, the 'S' displayed all the style, glamour and beauty that its older brother so obviously lacked.

At last MV had got it right, almost two whole decades since the first road-going shape of the ill-fated R19 prototype back in 1950. Gone at last was the 600's touring stance, replaced by a stunning creation, which had the crowds making a bee-line for the MV stand.

The newcomer was not only sleek, but also colourful, with its distinctive red, white and blue paint job. As always on a four-cylinder MV, the real centre of attention was the power unit, but this was now complemented by real street credibility – a bankable café-racer style, with clip-ons, rear sets, bum-stop racing saddle (finished in red), a jelly-mould tank, a four-pipe chrome-plated exhaust with matching megaphones, and massive Grimeca-made drum brakes, with a 220mm four leading shoe device at the front. Setting all this off was an abundance of highly polished chrome and stainless steel. The engine castings were left to a matt 'sand-cast' finish, as was traditional on MV products.

To this day, no one really knows what caused Count Domenico Agusta to make such a U-turn. Was it lack of sales of the 600, the desire for a machine that would justify the badge, or simply common sense? The truth behind the authorization of the 750S, coded at the factory as Type 214, will never fully be known, as Domenico Agusta died fifteen months later, in February 1971. Perhaps he had been inspired by an event that had taken place almost half-way around the world at the Tokyo Show in October 1968, in the shape of the launch of Honda's trend-setting CB750 four. In 1965, MV Agusta had had the distinction of being the first factory to present a modern across-the-frame, four-cylinder motorcycle – its 600 – but now Honda had upped the anti with the launch that really mattered commercially.

Sales of the CB750 exceeded 61,000 during the first three years in the USA alone, while the total production of four-cylinder MVs during the 1960s and 1970s is unlikely to have reached the 2,000 mark. Comparing Honda with MV was rather like comparing Ford with Ferrari.

1950s GRAND PRIX TECHNOLOGY

Like the 600, the 750S employed much the same technology as the 1950s four-cylinder factory racers. In practice, this meant a plethora of expensive gears, bearings and shims, all of which had the disadvantage for a series production engine of consuming large amounts of valuable build time. This meant that the finished motorcycle was not only too expensive, except for a handful of super-rich enthusiasts, but even this was not enough to provide the maker with a reasonable return on its capital. In contrast, Honda's product was far cheaper – in terms of manufacturing costs, if not in terms of quality – and made a far bigger profit per unit than each MV roadster four that rolled out of the Cascina Costa factory.

The harsh commercial fact was that MV had little chance of trying to beat the Japanese at their own game in producing a four that was cheap enough to build. Even if it had, it seems highly unlikely that they would have fared any better than Benelli, the one Italian marque that did. Benelli (then part of the De Tomaso empire) did have the advantage of starting with a clean piece of paper in designing its four- and six-cylinder models, which the MV engineers did not. (And the Benelli multis borrowed heavily from Japanese (read 'Honda') engineering concepts.) However, even with this advantage, Mr De Tomaso was to discover to his cost that it was not good policy for a small-volume Italian bike builder to attempt a head-on clash with the Japanese!

Perhaps what MV and Benelli should have done (with the advantage of hindsight) is to take the individual route, like Ducati, Laverda and Moto Guzzi, designing a new, simpler, cheaper, easier-to-build machine from scratch. For example, Moto Guzzi's almost agricultural automobile-type engineering produced a pushrod V-twin with the advantages of shaft final drive, which exemplified the ideals of simplicity in mass production and commercial profitability. Of course, MV really wanted to produce a motorcycle that was clearly influenced by its Grand Prix racing successes, and would mirror the two main

aspects of the factory's GP racers – publicity and glamour. However, the other MV roadsters of the period may have been humble ohv single and twins, but they did have one huge advantage over the fours: they made a profit!

THE ENGINE

Although the new 750 might have had an entirely new style, it actually owed far more to the 600 than might have been supposed. The engine layout, in particular, was really quite similar, and both had much in common with their racing brothers. As a first step in the transformation process from gawky 600 to svelte 750, MV's engineering squad increased the bore size to 65mm. However, the stroke remained as before at 56mm, this simple change not only giving over 150 more cubes – 743cc compared with 591.8cc – but it also made for a much shorter stroke. The engine was now considerably oversquare.

The pistons, three-ring Borgo-made assemblies (as on the 600), featured a 9.5:1 compression ratio (compared with 9.3:1 on the 600). The 30mm diameter inlet valves were the same as on the earlier roadster four, but larger (29mm) exhaust valves were employed. Perhaps surprisingly, the 750S also used the 600's camshaft profile, with its 8mm of lift. Carburation was also the same, with 24mm Dell'Orto UB24 instruments, but this time one for each cylinder, rather than one for a pair of cylinders on the smaller displacement model. These featured side-mounted float chambers, and were mounted parallel to one another. They were equipped with long, angled, polished aluminium bellmouths, with the inner pair angled outwards and the outward ones inwards.

With this level of tuning, the 750S offered maximum power of 65bhp at 7,900rpm to its rider. However, in this particular case, outright power was only half the story. Unlike the 600, which tended to be something of a buzz-box (and therefore unsuitable for its touring role), the bigger MV four had engine flexibility that was outstanding for a multi-cylinder machine at the time of its launch. In addition, its relatively mild state of tune ensured a level of smoothness not found on the later, more highly stressed models.

The official factory torque figures for the 750S were quoted at 5.9kgm at 7,500rpm, and to cope with the extra displacement and power output, the clutch was provided with fourteen plates, with the driven and drive ones being shared equally. As on the final batches of 600s, the gearbox featured a mainshaft with fixed (integral) first and second gears. The gear ratios were changed from those on the 600, becoming the following: 1st, 30/15; 2nd, 27/19; 3rd, 24/22; 4th, 22/24; and 5th, 21/25.

LIMITATIONS AND PROBLEMS

Much of the remainder of the power train was exactly as on the smaller four. This meant shaft final drive, which, for a motor-cycle with sporty pretensions, was something of a problem. The factory racers, except for the very earliest examples, had employed a chain, but for street use on the touring 600 Agusta the more civilized drive shaft had been chosen. The 750S retained this feature, which was, ultimately, to prove the model's Achilles heel when it was subjected to sustained hard riding. As events were to prove (*see* this chapter, and also Chapter 6), the choice of the shaft final drive was to limit the

(Above) Nearside...

...and offside views of the 750S engine assembly, mighty impressive from either side. Both photographs show post-1972 models with square-slide Dell'Orto VHB carbs

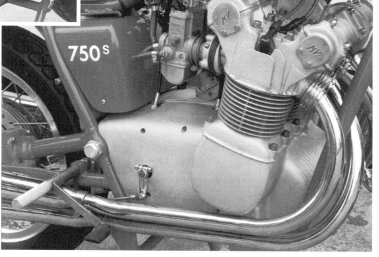

race potential of the 750. This limitation was to lead to its failure to offer much of a challenge in the prestigious Formula 750 category during the early 1970s.

Another drawback for road riding on the street was, ironically, the direct result of using a racing engine in a roadster. In its original guise, the design had had no provision for a generator, or, for that matter, for any form of starting device – except run and bump! Instead, the priority had been to keep the width down to a bare minimum, which in turn meant that it was impossible to mount anything at the end of the crankshaft. The MV engineering team was forced to come up with something else, and they introduced an externally mounted dual-pulley starter motor/dynamo below; this assembly was run by a pair of rubber belts behind the engine.

The result of these additions to an already heavy engine design, not helped by the employment of sand-cast crankcases and other engine covers (necessary for such small production runs), ensured that the 750S weighed in at a fairly hefty 505lb (230kg). With the factory four-cylinder racing machines, the potential weight problems had largely been solved by the extensive use of much lighter magnesium alloy. This was simply neither practical nor cost-effective on a production roadster, so the aluminium and its weight penalty had to be accepted.

Another problem was mechanical noise, which was much greater in the 750S than in the 600. Everything, aside from small end bushes, moved or rotated on either needles, rollers or ball bearings. This was not simply because MV Agusta wanted the power unit to be ultra-reliable, but was more a reflection of the design's age and inheritance. In the year when the four was first laid down, long before the days of high-pressure oiling systems and modern lubricants, there was no chance of the design team using plain

bearings. Their survival in what had originally been conceived as a racing dohc unit would have depended on a copious supply of lubricating oil. Without a totally new engine, the 750S was stuck with its racing origins – which did not include plain bearings.

When the rider fired the 750S's starter-motor button, the engine literally roared into life. With all those gears and bearings whizzing around inside, it was hardly surprising that journalists would describe the motor as noisy. American magazine *Cycle* commented as follows in its September 1973 issue:

> The churning gear noises under the tank sound like the world's most sophisticated gravel-crusher. But a touch of the throttle, the revs rocket, and the long tapered mufflers spit out a hoarse snarl. The engine responds so quickly that the twist-grip operates more like a stereo set's volume control than anything else.

STYLING

Unlike its smaller brother, the 600, the 750S was built for the sports rider, someone who lusted after the machine for its must-have-it-at-all-costs style, racing heritage, and the most coveted marque badge in the two-wheel kingdom. Of all the road-going four-cylinder models built by the old Meccanica Verghera concern, surely the 750S was the most pleasing to the eye. It was truly a beautiful creation and unmatched until the late 1990s and the arrival of the new F4 (*see* Chapter 14). For once, MV had built a roadster that could be admired, and would inspire craving in a way that only a handful of other Italian classic bikes of the 1970s could. This select band included the Ducati 750SS (round-case model), the Laverda 750 SFC and Moto Guzzi's original Mark 1 Le Mans.

Like the F4 of today, the 750S was a mass of small details, several of which were unique to this motorcycle, and which combined to give just the right effect.

At the top of the styling tree on the 'S' spec list was a superbly sculptured 5.28-gallon (24-litre) tank, followed by a racing seat upholstered in red leatherette, and four bright chrome-plated exhaust pipes with long, slow-taper megaphones. The stylists chose to angle the matching Veglia speedo and tachometers (later superseded by British Smiths instruments), and to encase them in polished aluminium pods. Tommaselli was chosen to supply the clip-on handlebars and controls. The air of quality was extended by the standard fitment of polished stainless-steel mudguards, Borrani welled-type alloy rims, massive drum brakes, Ceriani front forks (35mm diameter stanchions and exposed, chrome spring shocks). CEV provided the lighting equipment in the shape of a chromed 170mm headlamp and alloy oblong-bodied rear light (later batches came with a round unit, as fitted to Guzzi's V7S).

Paintwork comprised an Italian racing red for the frame, swinging arm, toolbox/battery box and the middle section of the fuel tank. The balance of the tank was in royal blue, except for a white stripe, which became larger towards the rear, and toward the front also. In the centre of this white line was the famous MV gear-cog logo.

PERFORMANCE AND IMPRESSIONS

The result, viewed from every which way, was a truly impressive piece of kit that looked to be travelling at a hundred miles an hour even when stationary. However, the truth was that the styling and the engineering qualities of the package were not matched by its performance. In its early brochure, the factory claimed 129mph (225km/h), but this simply was not true. When testing an early example, *Cycle World* achieved 114.06mph (182.5km/h) – with the tachometer registering almost 7,200rpm. However, because of the hand-built nature of the engine, individual units differed considerably. Even so, it is highly unlikely that a stock 'out-of-the-crate' bike back in the early 1970s would have achieved 120mph, let alone another 10mph on top of that.

None of the testers of the day was particularly impressed by the acceleration, either. The combination of a light flywheel and heavy dry weight (505lb/230kg) conspired to knock the edge off the machine's potential for get-up-and-go. Again, there was a plus side – this time, it was the engine's smoothness. The factory claimed 14 seconds for the standing quarter-mile. *Cycle World* got 14.5, with a terminal speed of 94.73mph (151.5km/h).

I recall an all-too-short ride in the summer of 1983 aboard a mint, low-mileage 750S, provided by Morini specialist dealer Elby Moto of Upminster, Essex. This machine (on sale for £3,000!) had been registered some ten years before, and had been loaned to me by owner Jon Green, to allow me test it in preparation for an article in *Motorcycle Enthusiast*. (Coincidentally, when I became editor of the magazine in October 1983, the front-page feature of that month's issue was none other than the very same 750S.)

My impression of the bike was that it was unlike virtually anything else on two wheels (except possibly the top-of-the-range Harley-Davidson V-twins). The experience was unique; the rider was transported to a different level, and made to feel really special. There was certainly a pronounced 'feel-good factor'.

Another instantly noticeable aspect of the 750S was that its racy riding position was also comfortable. In styling the bike, it

(Above) *Five-hundred four-cylinder Grand Prix racer of the type used between 1956 and 1965 by the likes of Surtees, Hartle, Hocking, Hailwood and Agostini.*

The chain final-drive four racer can be distinguished from its shaft predecessor mainly by the crankcase, which differed due to the transverse location of the gearbox and clutch. Note magneto at rear of cylinders.

The rarest MV Agusta racing motorcycle of all, the 348.8cc six-cylinder model of 1969. FIM rule changes meant that it was never raced.

(Below) *Making its debut at the 1969 Milan Show, the 750S entered production in 1971. This 1973 example has optional factory fairing fitted.*

Phil Read (2) rounds Mearside Hairpin, Olivers Mount, Scarborough, on his 500 four-cylinder MV, c. 1974.

(Right) *Phil Read on the start line with one of the wire-wheeled (1973 model) fours.*

(Below) *Giacomo Agostini's 350 GP bike of 1976, backed by the Api oil company, in MV's last year of racing.*

(Above) *Arturo Magni, whose MV career spanned the period 1950 to 1976, during which he held a number of posts including chief mechanic and team manager.*

Magni chain-drive conversion is a worthwhile modification for the 1960s/70s roadster fours.

(Below) *An 832 (862cc) Monza, c. 1977, with side panels and optional fairing removed to show frame and engine details.*

(Right) *Mark Kay (son of Dave) working on the top end of a 750S engine.*

Roland Brown testing one of the Magni-prepared four-cylinder specials.

(Below) *A Magni-MV four still looks like it's doing 100mph when standing still.*

(Above) *The Kay-built MV-based Ferrari motorcycle. Constructed in the early 1990s, it is the only machine of its kind.*

Dave Kay (left) and son Mark with one of the Kay MV replicas, c. 1994.

(Below) *Introduced in 1998, the 125 Planet is a typical modern Cagiva. It oozes style, and is an excellent motorcycle to boot.*

(Above) *The new MV F4 on display at the Centennial TT, Assen, May 1998.*

The F4 makes its British debut on the Three Cross Imports stand at the NEC Show, Birmingham, November 1998.

The author having some fun with the mini F4, May 1999.

(Below) *The legendary MV Agusta gear-cog emblem.*

(Bottom) *Mick Walker during his test session with an F4 Serie Oro, May 1999.*

750S (1971)

The 750S – at last, the company managed to create a truly sporting machine to capture the aura of the works GP racers, if not the performance. The first example was shown at Milan in 1969, but production did not get under way until 1971.

Engine

Type	Air-cooled dohc across-the-frame four-cylinder, with gear-driven cams
Bore and stroke	65 × 56mm
Capacity	742.9cc
Compression ratio	9.5:1
Carburation	Four Dell'Orto UB24B2 or UB24BS2 carbs
Lubrication	Wet sump, gear pump
Max. power (at crank)	65bhp @ 8,500rpm
Fuel tank capacity	24 litres (5.28 gallons)

Transmission

Gearbox	Five speeds
Primary drive	Gear
Final drive	Shaft
Ignition	Battery/coil, 12 volt

Frame

Tubular steel, full duplex

Suspension and steering

Suspension	front	358mm Ceriani teledraulic
	rear	Twin shock, exposed springs
Wheels		18in wire with Borrani welled alloy rims
Tyres	front	3.50 × 18
	rear	4.0 0 × 18

Brakes

	front	4LS 230mm drum
	rear	SLS 200mm drum

Dimensions

Length	84.5in (2,110mm)
Wheelbase	55.5in (1,390mm)
Ground clearance	6in (150mm)
Dry weight	505lb (230kg)

Performance

Top speed	125mph (201km/h)

seemed that the designers had not forgotten the rider, and the positioning of the clip-ons and rear set was almost perfect. There was a subtle 'kink' in the clip-ons, while the foot controls and a surprisingly luxurious saddle combined to provide the rider with a level of comfort far removed from the typical café racer ride of the period. It was less ideal in urban 'crawl' conditions, but at any speeds above 40mph (65km/h), the forward

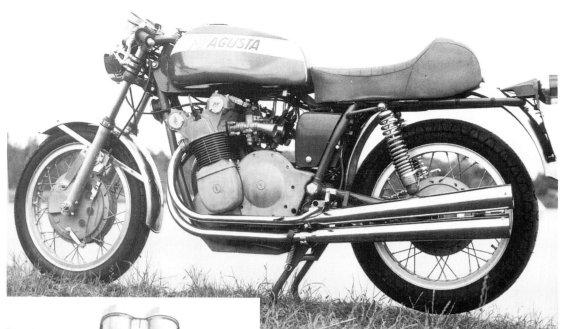

(Above) *From every angle, the 750S is an impressive sight.*

Narrow frontal angle, even though the 750S is a four.

(Right) *Frontal view of engine; main features include one-piece cylinder head, four barrels, horizontally split crankcases and massive build.*

(Above) *Engine casting – matt finish, but generally smooth.*

(Above right) *UBF round-slide carbs of original series 750S; note downwards-pointing bellmouths; distributor at rear of cylinders.*

Two instruments, two (Aprilia) handlebar switches, two warning lights and a centrally placed ignition switch.

Double-sided Grimeca drum front stopper – again from early series machines.

Shaft final drive, a feature not without its problems for 750S owners.

crouch allowed by the raised clip-ons was effectively counterbalanced by wind pressure to relieve any strain on the rider's wrists.

Considering how different the purpose and style of the 600 and 750 were, it is surprising how many of the cycle parts were the same. Perhaps most surprising of all was the fact that the two models shared both the frame and swinging arm. In fact, on the first 750S machines, the frames were actually resprayed modified components, left over from the 600 production run! These even retained the lugs (not used in this case however) for the forward footrest mountings and the bottom of the crashbars. Other identical parts included the rear wheel (and front rim), and all of the electrical system, except for the battery.

The 750 seemed to possess a notable improvement over the 600, suffering far less from that old bugbear of final shaft drive – the dreaded torque reaction.

PRODUCTION IN FULL SWING

By 1972, the 750S 'production' was in full swing, and more examples became generally available, with a number being exported to markets including North America, Great Britain, Germany and Australia. In the USA, the importer was Commerce Overseas Corporation, an organization operating from New York, which listed the 750S at 4,500 US dollars. In Britain, Vincent Davey was head of well-known London dealers Gus Kuhn, which were appointed sole UK concessionaires in October 1972, and listed the 'S' at £2,300. For the Australian market, the model was handled, together with other MVs, by the Bob Jane Corporation, appointed following a flag-waving racing visit by none other than multiple World Champion Giacomo Agostini. With depots in Adelaide, Brisbane, Footscray,

Granville, Hobart and Perth, the Jane Corporation effectively covered this vast continent. When the Australian magazine *Two Wheels* road-tested a 750S in October 1972, its price was quoted at 3,889 Australian dollars.

REVISIONS

Towards the end of 1973, a revised version of the 750S was shown, although production did not commence until early 1974. This updated machine, beginning with engine number 2140418 (Model 214, 418th constructed), featured alterations to both the state of engine tune and to the braking specification. In an attempt to improve the performance, the engine's camshaft profile was changed for a more sporting one, with 8.5mm of lift. There were higher (10:1) compression larger valves (31.8mm/27mm) and Dell'Orto square-slide VHB carbs, each piston equipped with its own float chamber.

The result of these changes added an additional 4bhp, making 69 horses in all. At the same time, maximum revs increased to 8,500. However, the downside to these modifications was that the engine became less flexible and more vibratory.

The cylinder head was totally re-cast to take into account the above changes. The combustion chamber shape was also altered. Externally, this type of head is noticeable because it has only a single horizontal bottom cooling fin, compared with two on the earlier 750/600 engines.

Another change was made to the clutch gear, which was reduced in width from 24mm to 17mm, thus allowing the clutch-cover face to be at 90 degrees to the camshaft, not angled as on earlier engines. This thinner gear can be removed and replaced with the mainshaft *in situ*, and the new casting had an edge bead to allow mismatch-mating to the crankcase, thus reducing the cost of hand fettling the castings.

From late 1972 (from engine number 2140418) an improved 750S was offered. There were alterations to both the state of engine tune and to the braking system.

The bike was now capable of exceeding 120mph (190km/h) and, probably because of that extra speed, the massive Grimeca front drum brake was ditched in favour of a more modern disc set-up – twin 280mm cast-iron assemblies with twin-piston Scarab calipers.

The colour scheme was largely the same, but, from the end of 1972, the white line on the tank had become the same thickness throughout its entire length, with bold 'MV Agusta' lettering superseding the original MV gear-cog emblem. Finally, an optional extra curved perspex screen was made available.

POLITICS AND MANAGEMENT

Behind the scenes there were a number of reasons for the updated 750S – and, perhaps more importantly, for subsequent policy within the two-wheel division of the Agusta empire.

The first was the sudden death after a heart attack of Count Domenico Agusta in Milan in February 1971. Although his younger brother Corrado succeeded him, he was never to enjoy the total control exercised by the autocratic Domenico. Such control was impossible in any case by 1973, as

the Italian government, in the shape of its financial institution EFIM, had taken control of Agusta. This was not because the group was in any sort of financial trouble, but because the government decided that it needed to preserve national interests concerning military contracts in the aviation side of the group. By now, this had become the dominant sector in the organization.

By the end of 1973, a new general director had been appointed by EFIM to head up MV's bike operation. Count Corrado Agusta, although still retaining the honorary title of President, was now principally committed in an executive capacity to the aviation arm.

The new man in the hot seat was Luigi Ghisleri, who was given the task of turning MV's racing successes into a commercial profit via the standard production range. This was no easy task, given that he also had to cope with severe financial restrictions placed on him by his government masters.

Ghisleri's first move following his appointment was to call a press conference. At this event, he talked about developing MV in a similar way to Ferrari in the four-wheel world. At the time, many of those present asked the question: Was this a high-flying dream, or was it sound commercial judgement? On that day, no one, not even

Ghisleri, really knew what the answer would be, but Ghisleri obviously believed he could turn things around.

At a press conference almost exactly twelve months later, in November 1974, Ghisleri made it quite clear that no change would be made to his policy. He also stated that MV not only intended to remain in Grand Prix racing, but also to be represented in the emerging Superbike category. He saw the company 'adding a luxury note to the smaller production classes' (meaning 125 and 350cc). He also pointed out that his presence confirmed that MV Agusta was being run by a government-appointed official, rather than by a member of the Agusta family; it was something of a slap-down for Corrado Agusta.

Italian journalists, in particular, asked why MV had not made more progress during the year. Ghisleri replied by pointing out that the 750 had been updated ('improved' was his actual word), and that production of the new 350 (a pushrod twin with square engine covers) had been held up by the disruptive action of trade unions. This last comment referred to the months of protracted negotiations, which had finally been settled by a new pay and conditions agreement.

A note of warning was also sounded in two comments made by Ghisleri: 'With the present world-wide economic crisis [the Arab oil embargo and its disastrous effect on world trade] we have to consider the market carefully for the expensive MV machine', and, 'We do not make high profits, believe me. In future we will have to make great efforts to justify the high cost of our machines compared to our competitors.' This was a clear message, if one was needed, that Ghisleri's government taskmasters were applying pressure regarding the future of the MV Agusta brand name as a motorcycle manufacturer.

Luigi Ghisleri gave his assembled guests a vision of the way he thought the top-of-the-range four-cylinder line should develop: 'In our opinion, the ideal was a sporting model. Now we are considering a change to a high-performance touring machine. The answer for the current market is a basic motorcycle that can be converted easily to race specification.' This statement showed that MV's

A display of MVs in January 1973 from the recently appointed British importers, Gus Kuhn: 750GT (left), 350B (top) and 750S (middle). The other bike is an 850 Norton Commando Interstate.

750 Super Sport

One of the rarest, if not the rarest of all four-cylinder MVs that actually made it to production was the 750 Super Sport, which debuted at the Milan Show in late 1971. Appearing exactly two years after the launch of the 750S, the Super Sport was in essence a show bike for the ultra-rich (costing some 30 per cent more than the already expensive standard bike). It was an undeniably attractive motorcycle. The Milan bike was in effect the sole prototype when it appeared.

Besides exciting the press and show-goers alike, this original prototype gave birth to a truly exclusive model, eventually to be built in very small numbers, for export to Australia (where it cost $4,779) and Germany. It also helped spawn the

The 1971 Milan Show saw the debut of the limited-production 750 Super Sport, with full fairing, racing tank and seat, Fontana front brake (a Grimeca on actual production models) and square-slide VHB Dell'Ortos.

factory F750 racer which Agostini rode at Imola in April 1972.

Features of the Super Sport included a revised fuel tank and saddle, and a full fairing (all in fibreglass). The fairing was later offered as an optional extra on the 750S. To improve the braking, a Fontana-made four leading shoe

(Left) The 750 Super Sport was the first MV production model to sport VHB carbs. The engine was specially tested by Agostini and bikes were sold with a certificate to prove it.

front stopper was fitted, together with British Dunlop TT100 tyres. Another feature was the use of four 27mm square-slide Dell'Orto VHB carburettors and a German Krober electronic tachometer. The seat was upholstered in white, instead of the red on the standard model. The buyer also received a certificate to say that his bike had been personally tested by Giacomo Agostini and that it had a specially tuned engine; it was a clever ploy to homologate Ago's Imola race bike, rather than simply offering paying customers more speed.

Today, the 1972 750 Super Sport remains probably the most sought after of all MV's so-called production four-cylinder models. This distinction is unlikely to change in the foreseeable future (as long as the new F4 series is discounted).

750 GT

Spring 1972 heralded another variant of the four-cylinder 750 MV street bike, in the shape of the now very rare GT.

Many informed observers insist that this machine was created in order to shift stocks of unwanted 600 parts that were gathering dust in the factory's

(Left) The first series of the 750 GT came with double-sided Grimeca drum front brakes. Note the curvy seat base. This particular motorcycle was sold by Gus Kuhn in 1973.

This 750 GT was purchased by MV enthusiast Mark Wellings in the early 1980s for the princely sum of £300. It had stood, seized up, against a wall for several years. He restored it to concourse condition in the mid-1980s.

Later version of MV's touring 750 GT with double-disc front brake and Scarab hydraulics on show in Amsterdam in 1974.

head man did not grasp just why people bought an MV, and gave a conflicting analysis of the problems the company was facing.

This muddled thinking also extended to a reply Ghisleri made regarding MV's production targets for the following twelve months parts warehouse, including crashbars, horns, handlebars and frames. Perhaps this is as good a reason as any.

Mechanically, the GT (Gran Turismo) differed from the 'S' in only two areas – the choice of compression ratio and the final drive gearing. As if to prove the theory that this was, in fact, a 'moving-stocks' creation, it was in many ways badly thought out. Not only did the mechanical items clash, but so too did much of the balance of the bike. Not only did it have a slightly higher compression ratio (9.7:1), but it was also given lower overall gearing – both features that were contradictory to its intended role. It was also fatally flawed as a viable touring machine because of a complete lack of any air-filtration system, a dual-seat saddle with a one-off pressed steel base (hand-beaten in pre-production prototypes) that was hard and uncomfortable thanks to an almost total absence of adequate padding, and foot controls that were positioned too far forward for optimum comfort. A couple of other noteworthy departures from the other 750s were a larger tool/battery box, and rubber front-fork gaiters.

Its paint job – a combination of a metallic golden brown and cream finish – was best described as flamboyant, and contrasted starkly with the 600's sombre, all-black finish.

Combined with such distinctive colours were the bulky slab-sided tank, high and wide handlebars, dual wind-tone horns and chrome-plated front crashbars. These components did give the right impression for a GT image, but the illusion was soon shattered by the machine's shortcomings.

The last say in the saga of the 750 GT must go to *Cycle* magazine, which asked the obvious question: 'Who would buy such a "touring" machine?' Even so, only fifty were built, and the 750 GT's rarity today means that it now commands a premium price.

total production over some five years of the 750S is unlikely to have exceeded 700.

Luigi Ghisleri's reference to the 500cc class is proof that the company was at least considering the production of a 500 four-cylinder roadster, as many observers suspected, utilizing the chassis of the Model 216 350 Sport twin, which had debuted in prototype form at the 1973 Milan Show.

RACING

Another question centred on whether MV would field a machine at Daytona in 1975. Ghisleri replied with a firm 'no', but went on to add that 'if a 750cc class is introduced to the World Championship, then we [MV] will enter machines with the same enthusiasm that we are currently showing in the 500cc class'. Significantly, 1974 was to prove to be the final year in which the Gallerate factory won a world championship, when Englishman Phil Read took the 500cc crown.

Inaugural Imola 200

Even though MV Agusta was never to fulfil Luigi Ghisleri's dreams (for that is what they proved to be!) of fielding an entry in the 750 World Championship, it did build such a bike. Its best-known appearance came on 23 April 1972, when no less than Giacomo Agostini took part in the inaugural Imola 200. The race is remembered now in the history books as a legendary 1-2 victory for Ducati riders Paul Smart and Bruno Spaggiari, but it had been 'Ago' and his MV ride who had started the day as favourite. The MV name had been responsible for much of the pre-race hype, with the resultant large number of spectators and coverage by the world's press.

Formula 750 racing was responsible for gathering together probably the widest-ever range of marque and engine types for a

(1975); he predicted a figure of 10,000 machines a year, equally divided between 125, 350, 500 and 750cc models. In fact, the

The author testing a drum-braked 750S during his spell as editor of Motorcycle Enthusiast *in the mid-1980s.*

single race class, before being virtually ruined by the two-stroke dominance from 1973 onwards. Its finest year was probably 1972, when the Imola meeting attracted not only the MV and the Ducatis, but also BSA, Triumph triples and Norton twins, Moto Guzzi V-twins, Laverda parallel twins, BMW flat-twins, Honda fours, plus Kawasaki and Suzuki two-strokes – all as factory or works-supported entries.

The ensuing crescendo of noise and colour as the race got under way was a sight unmatched even to this day. Ranged against Ago, Smart and Spaggiari were riders of the calibre of Phil Read, Walter Villa, John Cooper, Peter Williams, Ray Pickrell and Jack Findlay. As if to prove the pundits right, Agostini led as the flag dropped, and retained the position for four laps, before having to concede best to the flying Ducati V-twins. Agostini refused to throw in the towel and clung on to third place, and within striking distance of the two leaders, until lap 42, when the four-cylinder MV retired at the Acque Minerali section of the course. His retirement was due to that fault that always plagued the produc-

tion fours – the set of bevel gears located in the gearbox cover behind fifth gear, which turned the drive though 90 degrees to mate up with the shaft.

Formula 750 Regulations

Formula 750 regulations stressed that the bike had to be production-based, so MV was forced to employ the original shaft drive, despite its known weakness. Obvious and allowed modifications from stock included the racing Fontana front brake (also found on the limited-run 750 Super Sport) and four open megaphones. The engine tuning allowed was more a case of special attention (blueprinting), than a radical redesign – again, the major assemblies had to be as on the production bike.

Although Ago was entered by MV for the 1973 Imola 200, he was actually a non-starter; instead, he was practising on his GP bike at another venue on the same day.

The 750F1 racer was subsequently modified by a chain-drive conversion. Another 750, this time private, was converted for track use by British importers, Gus Kuhn. Basically a stock 750S with a half fairing, 4-into-1 exhaust and racing tyres, it was raced at Silverstone in August 1974, in order to evaluate its potential. Rider Dave Potter was not impressed and the whole project was soon abandoned.

For many, the 750S remains as a lasting tribute to the long-gone original MV Agusta marque. It was the nearest most people could get to owning a machine so closely associated with such a charismatic Grand Prix history. Even with its faults, and the fact it was hardly the speed machine it appeared to be, the 750S allowed its pilot to fantasize about the legends of the Grand Prix circuit – men such as Graham, Surtees, Hocking, Hailwood, Agostini and Read – and that is why it is such a valuable prize today with collectors and enthusiasts alike.

8 The American Influence

THE FIRST CONNECTION

During the mid-1970s, when the Cascina Costa factory was being run by the Italian government-backed EFIM organization and its appointed man Luigi Ghisleri, MV began a project with an American connection. Unlike the Agusta-Bell helicopter deal, this marriage was not destined to achieve the same levels of success.

A connection with the aviation side of the business was established via two Americans, Jim Cotherman and Chris Garville, who were based in New York. Garville was the head of Commerce Overseas, a company with offices in Wall Street, which had for many years been associated with the helicopter arm of the Agusta empire. Cotherman owned a State-side MV dealership and had built up quite a reputation with MV buffs in North America as a man with a genuine interest in racing, and, perhaps more significantly, as a tuner of the MV Agusta four-cylinder engine.

Cotherman and Garville met and soon became good friends, and, through their mutual interest, Commerce Overseas began to import small quantities of both the 750S and 750 GT models (*see* Chapter 7) during 1974. The two Americans were interested in MV motorcycles on a broader front, and came to the conclusion that, with a number of changes and updates, the four-cylinder model would be more saleable in the US marketplace. With this in mind, in autumn 1974, Cotherman and Garville organized a visit to the Gallerate factory to discuss the possibility of producing 'an improved 750S'.

SPAIRANI AND THE AMERICA

Rather surprisingly for a company that, in the past, had been noted for its conservative

After the 750S (and including the GT and Super Sport) the next model in the production four cylinder series was the America. Two Americans, Jim Cotherman and Chris Garville, sold Agusta the idea during a 1975 visit.

approach to marketing its production road-ster range of motorcycles, MV took up the American proposal with considerable enthu-siasm. Work began almost immediately on a revamp of the old model. The authority and direction for the project was put almost entirely in the hands of one man – none other than Fredmano Spairani, formerly of Ducati, and one of the driving forces behind the birth of the Taglioni-designed bevel V-twin.

Obviously, his stint at Ducati had taught Spairani an aptitude for motorcycle produc-tion planning. He brought to MV Agusta the kind of decision-making that was so lacking in all Italian industries in the 1970s, where the general attitude was that *domani* ('tomor-row') would do – in other words, why do today what you could put off until tomorrow?

Spairani was not just concerned with the initial design or construction of the proto-type but also with the 'nitty-gritty' of the actual production process. As proof of his drive and enthusiasm, it was less than fifty days before the first prototype of the machine that was to emerge as the 750S America was being road-tested!

The name 'America' was chosen for sound commercial reasons. The two Americans, Cotherman and Garville, had sweet-talked their Italian hosts into a significant expen-diture, and making the project a high prior-ity for Spairani and his development team, by promising a very interesting level of pro-jected sales. The main market was intended to be the USA, which was potentially vast, and by the time Cotherman and Garville departed for home, the factory believed that the America could realize some 500 sales a year. With this in mind, production was geared at a higher level than had previous-ly been attempted with any of the Gallerate four-cylinder models.

To this day, exactly how many Americas were built remains one of motorcycling's unsolved mysteries, but, from the events that were to follow, it would seem that Cotherman and Garville's estimates were optimistic, to say the least. Suffice it to say that bikes built during 1975 were still being sold off from dealers' showrooms as late as 1979 – and many of these were outside the United States. By this time, the original tyres were practically rotting away. It is also a widely held view that many of the later Monza mod-els were actually converted Americas.

CHANGES AND UPGRADES

Engine

Even though the America had the '750' des-ignation, the actual engine displacement was 789.7cc, achieved by enlarging the diameter of the three-ring pistons by 2mm. This gave bore and stroke dimensions of 67 × 56mm. The Borgo-made pistons gave a compression ratio of 9.5:1 (the same as the 750S), but featured a wider squish band and higher dome. The combustion chamber shape was revamped and the camshaft lift was upped by 0.5mm to 8.5mm.

In spite of the revised combustion cham-ber shape, the America employed the same cylinder-head casting as the later 750S; the changes simply calling for re-machining. However, this left one glitch, caused by the selection of carburettor type. Although both the 750S and America utilized the square-slide VHB Dell'Ortos, the later 750S had been provided with 27mm instruments, while the larger bore engine of the America was given 26mm units, which did not match the inlet ports correctly.

Why did they change to 26mm assemblies at all? The truth is that the 27mm instru-ment suffered from not being sealed at the top where the cables passed through to the slides. Another vice was that they would go out of synchronization far too easily. The

Factory brochure for the America. Although originally planned mainly for export to the USA, many of the machines were actually sold in Europe, notably West Germany. The America was not a '750', instead it had a displacement of 789.7cc (67 × 56mm).

smaller type not only featured sealed tops, but also had the added advantage of allowing the fitting of a crossover shaft with four arms (done by MV engineers), one for each slide. This feature allowed a single operating cable to connect between the shaft and each carb, solving the nuisance suffered by the original 750S, in which the instruments would often go out of synchronization on a bumpy road!

It was also possible to mount the quartet of 26mm carburettors rigidly on a single aluminium plate and secure this to the engine with four rubber sleeves, whereas on the 750S the carbs were individually mounted. Not only did this new mounting serve to isolate the carburettors from vibrations, there was the added bonus that a single air-filter box could be fitted. This was an important innovation, for it allowed the factory, for the first time, to equip one of its production fours with a decent air-filtration system. This air box was a black plastic moulding (although pre-production prototypes featured an aluminium version), and contained a very simple inch-thick coarse foam filler element.

Exhaust Options

The factory offered two types of exhaust systems for the America, one fully homologated and one not. The latter was an almost straight-through set of bright chrome-plated shallow-taper seamed megaphones. The other was an extremely muted, but ugly, pair of stacked muffles on each side of the machine, finished in a matt crackle finish. This latter system reduced the noise level to a socially acceptable 85 decibels, but was universally loathed for its lack of spirit. After all, who wanted a sanitized-sounding motorcycle? Certainly no one buying an MV four. It might have been all right on a Japanese machine, but certainly not on a blood-red Italian fire engine!

Legislation

In order to meet American federal legislation, the gear-change lever was moved from the right to the left – Japanese-style. This change was neatly executed and concealed

An America with all the factory-fitted options – cast alloy wheels, triple disc brakes (the wire-wheel version only had double discs at the front, and a drum at the rear) and full fairing finished in the works livery.

behind the engine casings, with no unsightly external rods or joints to spoil the clean lines of the power unit. The crossover shaft was only visible from the rear of the machine.

Performance

The changes introduced on the America resulted in a 75bhp (crank) reading, at 8,500rpm. Maximum torque figures came out at 47.88lb/ft (6.162kg) at 7,500rpm. MV sources claimed a 135mph (215km/h) top speed, with in-gear figures of 1st 57mph (91km/h); 2nd 80mph (128km/h); 3rd 105mph (168km/h); 4th 125mph (200km/h).

Switchgear

Another important upgrade concerned the handlebar switchgear. Formerly, all four-cylinder MVs had suffered from the dreadful moped-style components that had been fitted. These looked totally out of place on such an exclusive (and expensive) piece of machinery.

The Aprilia electrical concern (no relation to the modern-day motorcycle builder of the same name) supplied MV (and virtually the rest of the Italian motorcycle industry) with its wares. The company ceased to exist at the end of the 1970s and, towards the time of its demise, it had begun to manufacture what was obviously an Italian-made copy of the Nippon-Denso switches, which were fitted as standard equipment to the likes of Honda and Suzuki, plus some Ducati and Laverda models. Like the Japanese assemblies, the Aprilia ones sported cast-aluminium bodies and black plastic buttons.

Braking Equipment

The calipers were identical assemblies to those of the 750S, with identical dual 280mm diameter cast-iron discs at the front. The rear brake, a 200mm full-width alloy drum, with a steel liner, was again from the later 750S, but for the American market the back plate was equipped with a pair of small circular rubber inspection plugs to allow the owner instant access for shoe wear. Both wheels had Borrani-made welled aluminium rims, with German Metzeler tyres, a 3.50 × 18 ribbed front, with a 4.00x18 block rear tread.

Running Gear

The basic running gear of the America came largely from the later version of the 750S, with minor detail and cosmetic changes, together with a new styling mould. The frame design followed that of the 750S, but was now painted silver, as were the fork yokes. The Ceriani-sourced front forks featured larger 38mm diameter stanchions (35mm on the 'S'),

A wire-wheeled 750S America, fitted with performance exhaust and full fairing. Note also non-standard mirrors.

while the two-piston brake calipers were carried at the rear of the fork sliders.

The Look

The fuel tank was a prominent feature on both the 750S and the America, but they were strikingly different. The America's tank was a bright red 5.2 gallon (23.64 litre) affair with broad silver bands on both sides flanked by a pair of very thin silver coach lines. Manufactured in steel, the tank offered squared, almost chiselled lines, as did the converter single/dual saddle, with suede leather upholstery. The seat base was manufactured in fibreglass. One feature of the design was a small lockable flap, which gave access to a storage compartment within the hump of the seat base.

As with the saddle, the side panels and front mudguard were made from red fibreglass. The rear mudguard was the same stainless-steel assembly found on the 750S, but now painted red to match the rest of the bodywork. The headlight (Aprilia) and several minor components were finished in black, and there were metal badges on the tank ('MV Agusta') and side panel ('750S America').

RESPONSES

The first news of the imminent launch of the America reached the UK in early 1975, when the 12 February issue of *Motor Cycle News* carried a story headlined 'Restyled 750 MV for America'. Stateside *Cycle*, in its May 1975 issue, got several of its facts wrong, including the engine displacement and compression ratio. Interestingly, the same journal reported low levels of vibration, saying that 'the engine vibrated less than a Honda CB750, but perhaps more than a CB550'. Its reporter, and others, were also impressed with the riding position for such a sporting machine. The secret was a near-perfect layout of the folding footrests and raised clip-on handlebars. This, together with a low (29in/74cm) seat height, was much praised.

Even so, city use was not the America's forte. Once out of town and on the open road, it really began to come into its own. At a dry 518lb (235kg), the machine could hardly be called a lightweight, or even a middleweight. *Cycle* summed up the steering as being 'slow and heavy', and went on to say that in their opinion the MV 'didn't handle as well as the Laverda 1000 triple, but was more stable than a competently prepared Kawasaki Z1'. *Cycle*'s tester achieved the standing quarter-mile in 13.6 seconds, with a terminal speed of 105.4mph (168.6km/h). Later, the British weekly newspaper, *The Motor Cycle*, recorded the lower figures of 13.8 seconds and 100.2mph (160.3km/h).

The America's launch price on the US market in early 1975 was US $6,000 landed at New York. In addition to this, the buyer had to add extra for variable state taxes and transportation costs; to the West Coast, this could be significant. All this made the MV the most expensive standard production bike ever to hit the US market at the time.

If the cost was spectacular, so too was the paint job. *Cycle* called it 'brilliant in red-and-silver', and went on to report that 'travelling in a mushroom cloud of sound, the MV is far more than a stylish flash of light and noise!'

OTHER MARKETS

In the end, pricing concerns triumphed over the appeal of style and the America achieved poor sales in what had been intended as its main market.

Recognizing that the America was not going to reach its projected sales targets in the USA, the MV management in Italy decided to explore other export markets: Australia, Britain and Germany. It is interesting to note that the factory had never intended to restrict its efforts to the USA alone. Proof of this exists in the machines themselves, on which detailed specification changes were accommodated: wiring looms, headlamp types (replaceable bulbs for non-US markets, sealed beams for the US, with a thick rubber gasket between the rim and shell of the headlamp). There were also two distinct types of direction flashers (indicators). The first was chrome (and changed to black); those with side reflectors were originally intended for US market machines.

British buyers had to wait for the 750S America, as Gus Kuhn had relinquished the import concession in 1975. The UK Ducati and Moto Guzzi concessionaires Coburn and Hughes of Luton had held secret talks at Cascina Costa in the spring of 1976, but nothing was to come of them. Instead, an entirely new company, Agusta Concessionaires (GB), was set up, under the aegis of a Slough-based Saab car dealership.

In early December 1976, the new British importers held a track day for the press and potential dealers at Mallory Park, Leicestershire. Unfortunately, they could not have chosen a worse time – that very week, the world's biking press was carrying 'End of MV' stories across its front pages, and there were even rumours that Ducati (also government controlled) was to merge with the Gallerate concern. None of this came to pass, but the stories did not help the new importers in their bid to sell bikes! Instead, they served to alert the world to MV's growing problems.

Agusta Concessionaires' spokesman, Peter Bate, tried to present a positive face, saying:

> Despite the high cost of the machines, which will make them the most expensive on the British market, it certainly seems that dealers feel confident of customers.

So you think you're ready for an MV Agusta?

The MV Agusta 750S has 37 World Championships behind it. If you think you can handle the power, mount one.

37 TIMES WORLD CHAMPIONS

A 1977 advertisement placed by the British importers, Agusta Concessionaires of Slough, Bucks. Note there is no mention of the 'America' name.

To Bate's credit, he and his team were successful in appointing a reasonable number of dealers during their reign of a little over two years, which was something that Gus Kuhn had never achieved.

The America cost a cool £3,187 when it was launched in December 1976. Buyers could also purchase one of the factory-made fibreglass full fairings, and cast-alloy wheels with triple 280mm discs. The wheels were to cause a number of problems for British owners. The factory-fitted kits included not only the new wheels but also three identical 280mm discs. There had never before been a disc on the back wheel, but Agusta Concessionaires assumed that the apparently identical 280mm front discs from the wire wheel would fit the front cast wheel properly. How wrong could they be? In fact, for the cast wheel to be fitted, a 0.5mm recess on the rear of the discs was needed. Without this, the whole disc would crack. At least three machines supplied by the importers suffered this fate, ultimately failing the British MOT (Ministry of Transport) test because their brakes were damaged. Worst of all, the bikes' owners were totally unaware of this potentially lethal fault.

Early Americas had wire wheels; from 1977, the customer had the choice of wire or cast alloy. Like all MV fours, the America was not cheap. When it went on sale in Britain in December 1976, it cost £3,187 – and cast-alloy wheels or a full fairing were costly extras.

750S America (1976)

Engine

Type	Air-cooled dohc across-the-frame four-cylinder, with gear-driven cams
Bore and stroke	67×56mm
Capacity	789.3cc
Compression ratio	10:1
Carburation	Four Dell'Orto VHB 26 carbs
Lubrication	Wet sump, gear pump
Max. power (at crank)	75bhp
Fuel tank capacity	4.2 gallons (19 litres)

Transmission

Gearbox	Five speeds
Primary drive	Gear
Final drive	Shaft
Ignition	Battery/coil, 12 volt

Frame

Tubular steel, full duplex

Suspension and steering

Suspension	front	38mm Ceriani teledraulic
	rear	Twin shock, exposed springs
Wheels		18in wire with Borrani welled alloy rims
Tyres	front	3.50×18
	rear	4.00×18

Brakes

	front	Twin 280mm cast-iron discs with twin-piston Scarab calipers
	rear	SLS 200mm drum

Dimensions

Length	84.5in (2,110mm)
Wheelbase	55.5in (1,390mm)
Ground clearance	6in (150mm)
Dry weight	526lb (240kg)

Performance

Top speed	131mph (210km/h)

JUST 'A BIKE FOR THE WEALTHY CONNOISSEUR'?

In many ways, the America is the best known of the original MV Agusta company's four-cylinder roadsters. This is for several reasons: its looks, the numbers built, and the fact it was extensively tested by the press. It also formed the basis for the high-performance Monza that followed.

It definitely looked much better with the chrome 'performance' exhaust, rather than

Leicestershire enthusiast/ dealer Bill Johnson with his 750 America painted in Ago colours. This machine was imported into Britain in the early 1980s by George Beale. It is pictured at an MV Club Cadwell Park track day, circa mid-1980s.

(Below) *A lockable seat flap is a feature of the America (and later Monza) models.*

the muted black silencers. The performance exhaust, however, provoked a mixed reaction, particularly from non-motorcyclists, as this extract from the December issue of *Motorcycle Sport* reveals:

> There was a different response from the [police] when the MV passed through a Leicestershire village. Edging forward in a traffic jam, the bike stopped behind a patrol car, the windows were down and a pair of beady eyes looked for the source of the din. Things looked black when a brawny arm beckoned the MV, which sounded louder than ever, to draw alongside; then a voice commanded, 'Keep blipping it – we've never seen one before.' It turned out that both crewmen were motorcyclists (Norton 650SS and 150MZ).

However, the final words on the America must go to a quote from *Motor Cycle*:

> The MV Agusta 750S America is a bike for the committed and wealthy connoisseur. Many Japanese machines can provide similar specification and even have better performance, but none can compare to the MV's quality of sight and sound. The MV is to be ridden and experienced. And with the America having thrown off its previous reputation for having poor detailing and finish, it has risen to being one of the true classic motorcycles.

9   Final Days

AN END TO MOTORCYCLE MANUFACTURE

By 1976, MV Agusta was up against it, both on the road and on the track. In racing, the marque's final world title (out of a record-breaking thirty-seven) had been achieved by Englishman Phil Read, who had led the 500cc category throughout 1974. Even though Read managed runner-up spot the following year, the writing was on the wall. By the start of the 1976 season, he had upped and gone, and his place had been taken by none other than Giacomo Agostini. Considering 'Ago' himself had quit MV three years earlier, this move might seem bizarre. In truth, the Italian superstar was not rejoining with full works support; instead, it was more a case of him forming his own race squad, with himself as rider. Besides MVs for the 350 and 500cc classes, he also had the use of a TZ750 Yamaha for races such as the American Daytona 200 Classic.

A revised 350 MV four-cylinder racer was prepared by long-serving Arturo Magini and his team of mechanics (*see* Chapter 5), but, although it proved fast and powerful, it was not the most reliable of machines, finishing only one Grand Prix (with a race victory at the Dutch TT). As for the 500 four, this was simply being steamrollered by the hordes of four-cylinder Japanese Suzuki and Yamaha two-strokes that now dominated the blue riband class. Even so, Agostini did win at the final round of the year, held over the demanding Nürburgring circuit in the Eiffel mountains in Germany. As events were to pan out, this was a historic victory – MV's last-ever Grand Prix race.

That final round was not quite the end of the racing story, because MV sanctioned Ago to ride the 350 at Brands Hatch (and Read the 500 at Cadwell Park) after the GP season had ended. Then, the end of the line had definitely been reached; the machines were mothballed and the race shop was closed.

The decision to bring to an end almost thirty years of continuous racing had been taken at board level earlier in 1976, when a group of new directors had been appointed. The plain fact was the newcomers had been put there to cut the costs of the overall motorcycle operation, which was running at a loss, in contrast to the aviation division, which was seeing an ever-increasing growth in profits.

Out went the men who understood – or, at least, had learned about – motorcycles, such as Luigi Ghisleri and Fredmano Spairani, and in came helicopter specialists and accountants. From that day onwards, the fate of MV Agusta as a motorcycle manufacturer was sealed. The new board of directors had one simple brief: to clear the decks for a total concentration on the far more lucrative aviation side of the Agusta business. The above events were prompted by a trio of notable failures within the two-wheel division. The first was the poor sales of the America model, which had bombed against predictions, leaving an excess stock of unsold bikes at the Cascina Costa. The second was the high cost and poor performance of the racing team, in comparison with previous seasons. Thirdly, there was the cost of the new Boxer flat four dohc four-valve-per-cylinder 500 GP racer project. Designed by ex-Ferrari engineer, Ing. Bocchi, this had so far recorded figures little better than the existing across-the-frame 500.

Yet another nail in MV's coffin came with the resignation of Ruggero Mazza, the man who had designed all the MV GP racers since the 350 triple of 1965. At 54 years of age, Mazza had decided to go into semi-retirement, and had accepted a technical assistant's post at the small-capacity motorcycle builder Aspes.

Then came the first press reports (*see* Chapter 8) that MV was in trouble, and the rumours – which turned out to be true – that the famous name might never again be seen on the world's race circuits. The press, probably fed by leaks from within the Agusta empire itself, had a field day. On 8 December 1976, *Motor Cycle News* displayed the headline 'MV: End of Race Bikes', going on to say that 'production of motorcycles at MV's Cascina Costa factory is being run down, and although nothing has been finalized, the government-controlled Ducati concern may well retain MV in name alone'.

MCN's information was not quite accurate. Yes, there would be no more racing entries, but the Ducati connection was misleading. At the time, both MV and Ducati came under the umbrella of Italian government departments and financial administration, and the government's financial arm, EFIM, did appoint directors to both companies, with a brief to stem the losses! However, they were never connected either from a commercial or a technical standing. In a twist of fate, though, the two were both taken over many years later by the same company – Cagiva (*see* Chapters 13 and 14).

During this final period in MV Agusta's original company structure, it is difficult to separate fact from fiction. Certainly, with hindsight, it is possible to see that the motorcycling press often put two and two together and got five – or sometimes even six! However, it is known that, prior to the appointment of the 'final supper' board of directors in 1976, several production projects had been either considered or authorized for development. Among these was a three-fifty dohc twin consisting of half a four; a 500cc four; and even a 1-litre four. Fredmano Spairani and his associates had understood the need to capitalize on the marque's legendary reputation. Another project that was axed centred on a plan to sell privateers a production version of the works 500 four-cylinder racing engine. A batch of twenty-five such engines had been considered necessary for production to get under way. Sadly, through lack of funds in the motorcycle division, none of the projects ever came to fruition.

Only one new machine was authorized and reached production, but in many ways this was not a new design at all, but simply an enlarged and more powerful version of the America. In fact, it was a sales trick to shift the remaining stock of bikes and parts already in the factory! It also allowed more time to sell off the smaller models, in the shape of the 350 twin and 125 single (both pushrod machines, with unit construction engines and five-speed gearboxes). No modifications were carried out from the beginning of 1977 onwards.

THE MONZA

Specification and Preparation

To justify its 'new model' status, the Monza's engine displacement was increased to 837cc, even though in typical MV fashion it was referred to as the 832. The increase in engine size was achieved by adding an additional 2mm to the size of the cylinder bores, taking these out to 69mm, with the stroke remaining unchanged at 56mm.

Another change was, for the first time, a twin-point Marelli distributor, which superseded the original single-point German Bosch unit employed on all the earlier production four-cylinder engines. This innovation

---

Monza 832 (1977)

Engine

Type	Air-cooled dohc across-the-frame four-cylinder, with gear-driven cams
Bore and stroke	69 × 56mm
Capacity	837cc
Compression ratio	9.9:1
Carburation	Four Dell'Orto VHB27 carbs
Lubrication	Wet sump, gear pump
Max. power (at crank)	85bhp @ 8,750rpm
Fuel tank capacity	4.2 gallons (19 litres)

Transmission

Gearbox	Five speeds
Primary drive	Gear
Final drive	Shaft
Ignition	Battery/coil, 12 volt

Frame
Tubular steel, full duplex

Suspension and steering

Suspension	front	38mm Ceriani teledraulic
	rear	Twin shock, exposed springs
Wheels		Six-spoke EFM cast alloy, with gold finish
Tyres	front	3.50 × 18
	rear	4.00 × 18

Brakes

	front	Twin 280mm cast-iron discs with twin-piston Brembo calipers
	rear	Single 280mm cast-iron disc with Brembo calliper

Dimensions

Length	84.5in (2,110mm)
Wheelbase	55.5in (1,390mm)
Ground clearance	6in (150mm)
Dry weight	526lb (240kg)

Performance

Top speed	145mph (230km/h)

Only two motorcycles were imported into Britain and sold as 'Boxers' before car manufacturer Ferrari forced importers Agusta Concessionaires to change the name back to Monza.

provided a superior dwell time to cope with the additional performance, which was now up to a claimed 85bhp at 8,750rpm. (Owners of Monzas have to be alert to the fact that timing the engine on the first set of contact breaker points can cause holed piston crowns. This is because this first set of points gives a spark that is 15 degrees more advanced than the second.)

Another change to the Monza's earlier engine specification was a newly created inlet cam, which, together with an America inlet cam as its exhaust, provided ten degrees of overlap. The model also reverted to the 27mm carb size (Dell'Orto VHB square-slide instruments) used on the late 750 Sport; this ensured that the carb/inlet ports blended correctly, which had not been the case on the America. Because of this, the Monza also reverted to the type of carburettor mounting employed on the 750S, with the identical problems associated with carb synchronization when negotiating roads with poor surfaces.

Finally, a small but none the less important technical point was the compression ratio. Official MV sources claimed 9.5:1; in practice, the true figure was higher, at 9.9:1.

Monzas built at the factory were fitted with six-spoke EFM cast-alloy wheels finished in a gold colour as standard. To these were fitted cast-iron discs, dual at the front, a single component at the rear. All three discs bit on twin-piston Brembo calipers. Prior to leaving the confines of the factory, each Monza would receive a special preparation, carried out by a member of the now disbanded race shop, headed by Arturo Magni. Incidentally, Magni had been the first MV employee, apart from management, to learn of the race team's fate. His sense of loyalty to MV meant that he carried on doing his best for the company, right up until the final closure. (In the meantime, he made plans and, when MV Agusta was finally closed, he set up his own company just outside Milan, in mid-1977, using the name of his son, Giovanni.)

The Monza's performance was proof of Magni's continued attention to detail. It had a 145mph (230km/h) speed potential (over 10mph, 15km/h, up on the best Americas),

Monza with factory fairing – plus mirrors. Note non-standard beading around fairing and screen.

(Below) *The ill-fated 'Boxer' logo.*

which, for 1977, was an impressive achievement for a roadster with a dry weight of 526lb (240kg).

Marketing and Testing

The Monza was originally marketed in Britain as the Boxer, the logo on the side panel having an upper-case X. This led to a threat of litigation from Ferrari, which had already registered the name for one of its prestige cars. It was, in any case, an odd name for the bike, particularly in view of the fact that MV had had an ill-fated Boxer racer, which did not even share the same engine layout. It is not clear why Agusta Concessionaires wanted to use the name, but it seems to have been a misguided and ill-researched choice.

Research shows that only two machines left the importer's warehouse with the Boxer logo. One was purchased by John Safe, a property developer and later Chairman of the MV Owners' Club of Great Britain. The other was a bike used by ex-factory

star Phil Read, which was, to quote Read at the time, 'on permanent loan from Agusta Concessionaires'.

Keen to publicize the MV brand, Agusta Concessionaires (appointed towards the end of 1976) were generous in providing road-test machines for the British press. Unfortunately, many of these tests resulted in over-hyped journalistic licence rather than factual, detailed reports that a prospective buyer, or even an existing owner, could use to learn more about the machines.

However, one test was just the reverse, being both down to earth, and sometimes critical. It appeared in *Motorcyclist Illustrated*. The tester began by saying that however much Japanese makes might lack in terms of road-holding and handling qualities, they had brought to large-capacity, high-performance bikes a 'certain calm and effortlessness that derives almost solely from the excellence of their engines and the relative silence that surrounds them'. In short, Japanese machines were unobtrusive. The same could not be said for the formidable Monza, which took to the road with all the subtlety of a Grand Prix racer.

The *Motorcyclist Illustrated* test report underlined the fact that the Monza was at its weakest coping with urban traffic, but its overall view was that, despite its price, the Monza represented real back-to-basics motorcycling with 'a strong frame, taut suspension, a powerful engine, superb brakes – and precious little else!' Of course, during the late 1970s the expression 'strong residuals' had not been coined, at least not for motor vehicles, but it is certainly one that would have been used about the Monza.

Genuine or Fake?

Very few bikes of its era command such high prices as the Monza today, and it now ranks alongside such exotica as the Brough Superior SS100 and the HRD Vincent Black Shadow. Of course, this level of value leads to problems for potential purchasers, and even current owners, because not all 'Monzas' are the genuine article. Sadly there are fakes out there to catch out the unwary. For starters, any Monzas that have Scarab brake calipers, wire wheels, or drum rear brakes are most certainly not the real McCoy. Furthermore, if Brembo calipers have been fitted to cover this up, a problem arises, because Brembos have their pads positioned higher to the mounting holes than those of the Scarab type. Thus, the front fork mounting lugs need to be in a different position if they are to locate with the disc correctly. If Brembo calipers are fitted to the earlier disc-braked MV fours, or to Monza replicas, without changing the fork sliders from the Scarab to the Brembo variety, the top third of the pad is not in contact with the disc, giving dangerously inefficient braking performance.

This view of the Monza's engine shows to advantage the massively finned bulge at the front of the crankcase. Note addition of all-black Magni after-market exhaust system.

To confuse potential buyers further, then and now, British importers Agusta Concessionaires listed not only the Monza, but also (in 1978) the Monza 861 Arturo Magni. The latter was not, in fact, an MV production model at all, but a special built by Magni's newly established company.

Agusta Concessionaires were also offering 'Mag wheels' in their April 1978 price list. In fact these were not magnesium, but simply cast-aluminium alloys.

MV ASSEMBLY IN BRITAIN?

Just before spring 1978, *Motor Cycle News* carried a story full of contentious material, with a headline reading, 'MV Agusta, Italy's most famous racing and road machines, may be assembled in Britain'. This centred around a rumour that Agusta Concessionaires had been offered the chance to buy the famous marque 'lock, stock and barrel'. It went on to report that 'they are considering the name at the moment and managing director Peter Bate says the money could be made available if they decided to buy'.

Whether the British importers were in a financial position to negotiate seriously is open to question. Certainly, a letter exists dated 27 April 1978, signed by Peter Bate himself, which states that 'Due to the recent sale of our parent company Haymill Motors Ltd, to Saab GB Ltd, the company has taken over temporary storage and warehouse facilities nearby. The address is Agusta Concessionaires (GB) Ltd, The Dairies, Moneyrow Green, Near Holyport, Maidenhead, Berkshire. We will be moving into these new premises over the weekend 29 April/1 May.'

There are two different ways of interpreting the situation. Either it could have been a genuine attempt (although probably a foolhardy one at the time) to save the MV name, or it was a ploy to generate enough confidence with the press and public to enable the importer's remaining stock to be disposed of via the conventional channels. The fact is that just six months later, in October 1978, Agusta Concessionaires closed its operation, selling the remaining stock of bikes and spares to nearby SGT Superbiking of Burnham, Buckinghamshire.

THE LAST RITES

At the same time, MV in Italy had all but ceased to exist. The motorcycle production lines had ground to a halt for the last time, and the space they occupied had been gobbled up by the helicopter manufacturing arm. From then, until the company's rebirth under Cagiva control some two decades later, MV

MV enthusiast Dave Kay at the MV Owners' Club Track Day at Cadwell Park, July 1984. His machine has a non-standard paint job and Magni exhaust.

Mini Bike

The Mini Bike was a completely new concept for MV – a child's motorcycle. Now highly prized by adult MV fans, this son of the MV racers caused a stir much larger than its tiny size when it was launched at the 1975 Milan Show. This wonder toy managed to exude all the glamour of the real thing, even though it was hardly a practical machine. Unlike the popular dirtbike-style children's motorcycles, the Mini Bike was pure fantasy for a wide-eyed youngster with an equally wide-eyed (and well-off!) parent.

The Mini Bike could be specified in one of two versions; the only difference being the choice of 8 or 10in wheels. Both versions were styled just like a replica of the legendary MV four-cylinder racers. Right from its quartet of black-painted megaphone silencers, which gave no hint of the meagreness of the 47.6cc bought-in Morini-Franco two-stroke engine, with its single-speed gearbox and automatic clutch, the tiny MV oozed style and panache.

The number sold remains a mystery, and how many of these survived beyond those first few rides on the tarmac – with either the bike's fairing or the youngster remaining unscarred – is also open to question. What is known is that today they are highly collectable.

At the time, it was generally accepted that the Mini Bike came about when works rider Phil Read wanted MV to build his young son, Pip, a replica of his own bike. Read's company, Phil Read International, imported a small number

The Mini Bike, available with 8 or 10in wheels. A child's bike for future champions ... or today's collectors.

into Britain, displaying examples at the Racing and Sporting Show at the Horticultural Halls in Westminster, London, in January 1976.

Agusta was to remain a dream for the vast majority. The special builders, such as Magni, Hansen, Kay and Bold, effectively kept the name alive. They were joined in their endeavours by the various owners' clubs, particularly the British one, which purchased the factory's remaining stock of spares in the late 1980s, after MV's last employee retired from his post as parts storeman.

As for MV Agusta itself, this most glamorous and successful name in the history of Grand Prix racing was effectively dead for many years, the victim of a combination of events throughout the 1970s: the death of Count Domenico Agusta; a lack of financial

planning and investment in new models; the intervention in the company by the Italian government through the aviation side of the business; the success of the helicopter division; and, most of all, the interference of countless unnamed officials to whom balance sheets meant more than a name, however famous and highly regarded.

The gloom was to be lifted when Cagiva had the foresight to purchase the brand name, and eventually used it to launch what is generally regarded as the finest production motorcycle of the twentieth century – the sensational F4. So, most important of all, the MV Agusta name lives on. Hooray!

Hansen 1100 Grand Prix

Germany was (and still is) a centre of enthusiasm for the four-cylinder MV Agusta family. During the late 1970s and early 1980s, the Hansen concern, based in Baden Baden, West Germany, made an attempt to fill the gap in the market left by the departure of models such as the America and Monza. The men behind the scheme were Michael Hansen (formerly MV's German importer) and partner Roland Schneider. Hansen tried to get Arturo Magni involved in the project, but the former MV race manager declined the offer. Undaunted, Hansen went ahead

The German Hansen 1100 GP of the early 1980s. Only eleven were built.

Post-closure MV fours, except Arturo Magni's own creations, are pretty rare. The 125bhp Hansen 1100 Grand Prix was powerful but had borderline reliability because of extra cubes.

manufacturing what turned out to be the largest-capacity version of the old air-cooled motor to reach (albeit strictly limited) production.

The centre of the 1100 Grand Prix was its 1066cc displacement, thanks to its 74 × 62mm

bore and stroke dimensions. To achieve this, it needed special items, such as cylinders, as well as a new crank carrier to allow the larger 74mm bore size. The larger pistons were manufactured by Marle and were of the forged type.

In appearance, the 1100 Grand Prix was much like the Monza, but with larger Dell'Orto carbs with accelerator pumps. The exhaust was supplied by Magni, wheels by Campagnolo or FPS and brakes by Brembo; front forks were usually by Forcella Italia (formally Ceriani), as were the rear shock absorbers.

Despite a massive price tag, Hansen found buyers willing to pay for the MV experience, but the machines were not always reliable. Indeed, Dave Kay claims that, of the eleven machines constructed, no less than seven experienced an engine blow-up. Even so, Hansen was, in many ways, 'ahead of the game'. He was, besides Magni, probably the first to offer customers something they could no longer buy from MV itself.

Finally, in 1985, the German magazine *Das Motorrad* carried a story, together with an artist's impression of a new 1100 l6-valve MV-based Superbike, which it was reported would be built in co-operation with Magni. This grand plan, reputed to be backed by some two million marks (then around £500,000), did not proceed; again, Arturo Magni vetoed it, probably reasoning that building his own specials powered by BMW and Moto Guzzi engines would be a better bet.

10   The Classic Scene

ORIGINS OF THE SCENE

Genuine motorcycle enthusiasts have always had something of a fascination for older machines. It may be fun to have a shining new road-burner, but bikes from a past era have their own story to tell, and offer a sense of nostalgia, which a brand-new vehicle cannot match. Over the course of the last two decades, the market has witnessed a complete transformation and the single word 'classic' has come to mean 'collectable'. Over the same period, prices have risen stratospherically, and crashed in equally dramatic fashion. What was once the preserve of specialist engineers, and amateurs lovingly rebuilding the dream bike of their youth, has developed out of all recognition. On the one hand, it is a cult, but there is also an internal market, in which sadly the potential for investment is often a greater motivation than a genuine passion for the motorcycle itself.

It all started during the latter part of the 1970s, when interest in British machinery of the 1940s, 50s and 60s flourished. Machines were available at a reasonable cost, and the generations that had been in their twenties during these periods used spare capital to indulge a need to recapture their youth. Before long, book and magazine publishers were producing material to cater for the classic scene, as it was beginning to be called.

East Midland Allied Press in Britain was the first to spot this developing market, and in March 1978 it launched *Classic Bike*, the

A corner of the MV race shop at Perno in 1977. The dust is beginning to settle and the machines themselves are no longer awaiting transport to a race circuit. They were now only glorious souvenirs.

125

The British MV Owners' Club was formed just as the Classic scene was beginning at the very end of the 1970s. The photograph above shows some of the early membership during 1980.

world's first specialist magazine catering for the new breed of enthusiast. *CB*'s first editor was Peter Watson and the inaugural issue contained articles on vintage racing, side-valve engines, Ducati singles, dustbin fairings, the famous designer Edward Turner and War Department motorcycles, as well as a test report on a Norton Dominator (a 500 Model 7). Published monthly, the magazine celebrated its 200th issue in September 1996 and remains one of the world's top-selling motorcycling monthly titles.

THE CHANGING MARKET

The demand for classic machinery, combined with the interest of journalists, saw the value of classic bikes climb steadily throughout the early 1980s, before finding a level in the middle of the decade. Looking back, this was without doubt the golden era of the classic motorcycle and its genuine followers, with the market in a healthy state, but with prices still relatively low, except for genuinely rare and exotic models.

As the 1980s came to an end, dealers and speculators who were active in the historic car trade began to turn their attention to the biking world, with the inevitable effect of unbalancing the previously stable market base. There was a surge in values generally, although the greatest rises were reserved for the prestige models, notably Brough Superiors, HRD Vincents and MV Agustas. In autumn 1989, examples were changing hands for upwards of £25,000 ($40,000), with some Broughs commanding even more.

With the end of the 1980s, the market moved rapidly away from its enthusiast base to an investment-oriented one. The era of the 'bank-vault' motorcycle had arrived. In Britain, for example, hundreds of motorcycles, notably British and Italian, were being exported to Japan, where the yen-rich could afford to pay often double the price that could be realized in Britain. Often, these machines ended up being sold in Tokyo antique shops! The classic craze was sweeping Japan to such an extent that not only were genuine classic bikes becoming much sought after, but there was also a booming business developing modern-day classic styling exercises. Models such as Yamaha's SRX600 and Honda's XBR500 (both single-cylinder four-strokes) were dressed up to look like a Manx Norton, AJS 7R, Ducati Desmo or MV Agusta 750S.

Every boom has its bust, and in autumn 1990, certain factors combined to put the classic motorcycle market into a steep decline. The worldwide recession felt in countries such as the USA, Great Britain and

The Replica Industry

Because four-cylinder MVs are so expensive, a new cottage industry of manufacturing replica parts, or even complete motorcycles, has sprung up. In Britain, this industry was originally aimed largely at single-cylinder classics such as the Manx Norton and Seeley G50 racer, but, in the 1990s, it spread to much more exotic machines, including Benelli, Gilera and MV Agusta fours. The Moto Guzzi factory even came up with a plan to build a small batch of its own 1957-type horizontal singles, although this project was axed before production could begin.

There are men, such as Dave Kay in Britain and Albert Bold in the USA, who offer their own versions of the MV theme – and tell everyone exactly what they are selling. Unfortunately, there are others who get rich by passing off a replica as the genuine article; with modern engineering techniques, some replicas are so good that it is often almost impossible to tell exactly what is being offered for sale.

A replica of a late 1950s MV four racer built in Italy from parts manufactured by British company Kay Engineering.

In Italy today, it is possible to buy almost anything – particularly for those who have loads of money. Certainly, a number of Italian businessmen have cottoned on to the fact that there are many wealthy enthusiasts and collectors, notably from Germany, Italy and the USA, who have the money to spend on obtaining what they think is their own piece of motorcycling history.

For almost two decades a museum dedicated to MV's achievements in both the racing and production fields was open to the public in the centre of Gallerate, only a few miles from Agusta's manufacturing base.

(Right) There were almost 200 vehicles and engine assemblies in the MV Museum, including this 1953 500 four of the type raced by the late, great Les Graham.

For some two decades a museum dedicated to the achievements, both on and off the track, of MV was to be found at Via Matteotti, in the centre of Gallerate, only a few kilometres from Agusta's manufacturing base. Opened in May 1977, the Museo Della Tecnica e Del Lavoro MV Agusta was no mere private collection, but an official Agusta Group project. The museum's caretaker (a former MV engineer) lived in a flat next door.

The display, with almost two hundred vehicles and engine assemblies, covered not only MV's motorcycles, but also its other engineering achievements, from both industrial and avia-

tion fields. There was even an experimental three-wheel car dating from the mid-1950s, powered by a 350 twin-cylinder engine, and the weird and wonderful 'Overcraft' (actually a hovercraft powered by a 300cc twin-cylinder motor), dating from 1960.

Besides the actual exhibits, there was much documentation, not just connected with the marque's GP successes and its standard production roadsters, but also MV's participation in off-road events, including motocross and long-distance

Approaching the building housing the museum; this sign was the only visible evidence of the treasure within.

trials – even scooter racing! In a passageway outside the main show areas, an exhibition in a number of glass showcases contained items such as technical drawings and mechanical components employed in MV motorcycle production.

A 96-page illustrated booklet was available to those visiting the museum. Besides listing all MV's GP victories, and providing photographs and technical specifications of many exhibits, it also listed the countries to which MV motorcycles were exported (no less than fifty-four), the total number of championships and race victories gained, and also a tribute to the six MV factory-contracted riders (Vincenzo Nencioni, Renato Magi, Les Graham, Ray Amm, Roberto Colombo and Angelo Bergamonti) who lost their lives in competitive events.

The museum finally closed for the last time in 1997.

Japan resulted in many potential buyers becoming extremely wary; in many cases, those who had speculated found that they had become seriously overstretched. As had happened in the classic car world, this period saw an excess of machines being unloaded on to a depressed market, with the result of a titanic drop in prices. In the twelve months from September 1990 to September 1991, HRD Vincents and MV Agustas (such as the 750S and America) lost on average £10,000 ($15,000) in value, with similar losses repeated for Broughs and the vast majority of other exotic machinery that had witnessed such dramatic rises during the boom years.

Although speculators and investors had their fingers burned, the effects of the bust for the genuine motorcycle enthusiast were very welcome. Bikes at affordable prices flooded on to the market once again – the large numbers of popular classics available meant that any potential buyers could be selective in their ultimate choice. By 1992/93, a full return to the enthusiast-driven market had been achieved, with prices generally stabilizing. Perhaps the golden rule is always to buy a motorcycle for love, not for profit!

MILLER'S

In 1994, *Miller's Classic Motorcycles Price Guide* was published for the first time. This annual book has quickly established a strong and loyal readership around the world. It acts as a unique guide to classic and collector's motorcycles, covering an extensive range, from veteran and vintage to post-war classics right up to the beginning of the 1980s.

Besides providing a price guide, *Miller's* also reflects which marques and models have been offered for sale in Great Britain during the previous year. The book includes a comprehensive listing of auctioneers and dealers which provide information including the likes

A happy band of MV Owners' Club members on Douglas promenade during the early 1980s.

of Brooks, Sotheby's, Verralls, and specialists such as the Italian Vintage Motorcycle Co. and Atlantic Motorcycles, as well as the most important private transactions. There are also appendices including a Directory of Museums, a Directory of Clubs, Bibliography, Glossary of Terms, and other information. The 2000 edition has sections on dirt bikes, speedway, specials, scooters, sidecars, police bikes, racing machines and memorabilia, plus, of course makes from ABC to Zenith, with a comprehensive section on MV Agusta.

SHOWS AND AUTOJUMBLES

Another facet of the classic scene is the show circuit. One of the earliest events took place at Belle Vue, Manchester, supported by *Classic Bike* (the venue was later changed to the Stafford Show Ground), followed later by the Bristol Classic Bike Show. The biggest European classic motorcycle event – the Utrecht 'Old Timers' show and auto fair – is

held twice annually in the Netherlands, in early March and mid-August. The exhibition is held in a modern trade complex in the centre of Utrecht and attracts enthusiasts from all around the world. There are now many smaller shows, usually held throughout the summer months, ranging from popular provincial ones to local club affairs.

Autojumbles have become a way of life for many owners and restorers of classic bikes. Old bikes, just like new ones, need parts to keep them running in serviceable condition. However, getting the right part for an old machine is not simply a case of walking into a local dealer and placing an order. Often, enthusiasts will spend considerable time and energy looking, but consider it part of the job of ownership of a beloved classic (and of MVs, in particular!).

Autojumbles can often prove invaluable, and cover the biggest area at many shows. It may take hours to search through all the stock, so an early arrival is recommended. Even those who go away empty-handed are sure to have met up with a fellow enthusiast and to have picked up some useful leads for the future.

The stock is likely to range from what appears to be simply old junk, to a complete bike needing some care and attention – and everything in between. There will be baskets of bits that purport to have come from one machine, but quite often have not. It will obviously be more difficult to find parts for something as specialized and valuable as an MV than, say, for a Triumph twin, but miracles do happen. Sometimes, the rarest items turn up – and at bargain basement prices.

RESTORATION

Restoration is a word that looms large in classic motorcycling arenas. In fact, some enthusiasts enjoy the restoration as much as, if not more than, riding the finished machine. Restoring any bike, let alone a four-cylinder MV, can be a daunting task for the inexperienced, so here are a few tips which will save you time and money – and help keep you sane!

Where to start? That depends on whether you already own a bike, or need to buy one. Of course, it is vital to have the finance available, not simply for what you would like, but for what you can afford. Affordability is all-important; without sufficient funds, no one, however enthusiastic, will be able to carry through a full restoration. It is also wise to set a limit on spending. As a general rule of thumb, make an estimate of time and cost, and then add on at least half as much again.

The next item on the agenda is ability, which will determine what you can do yourself and how much will need to be farmed out. Tools and equipment are also vital. Most enthusiasts do not possess a set of factory tools at a fully-equipped workshop, but access to a clean, warm, well-organized workplace is important. With such a valuable piece of machinery as a four-cylinder MV, it must also be secure.

Draw up a workplan. This is vital to a successful restoration. Without a plan, the task will be a nerve-racking experience, and you may lose patience with the project and sell the incomplete machine out of sheer frustration.

Before starting out on a restoration, it is a good idea to research the spare parts availability and costs for the motorcycle. For a classic MV, the three main ways of turning up bits will be the owners' club, specialist dealers, and sometimes, if you are lucky, autojumbles. You could also try advertising.

Workshop manuals, parts books, magazines and books are all useful sources of information and joining an owners' club provides useful contacts. At the outset, the task

of restoration may seem a daunting one for the novice, but nothing compares to the sense of pride and achievement felt when you have finally seen the project through and your very own classic stands before you in all its shining glory.

CLASSIC RACING

Along with the massive upsurge of interest in classic bikes there has been a growth in the sporting side, notably in pre-65 trials and motocross and road racing. The cut-off date for the dirtbike side is 1965, but road racers can often use bikes from the 1970s as well. Today, classic racing is a vibrant part

Former MV world champion John Surtees has been a leading light in the Classic racing movement, from its early days. Here, he is seen at Brands Hatch in 1981, with one of the fours that he raced during his career.

Dave Kay

Dave Kay was born in Nottingham on 19 December 1940. A life-long motorcycle enthusiast, he has not only built some of the finest MV-based motorcycles, but has also often been the centre of controversy with his forthright approach to life.

A youthful Kay began his two-wheel career on a Sunbeam (BSA) 250 ohv scooter, which was followed by a new Triumph Bonneville, and later more Triumphs and, later still, Vincents. He first became the owner of an MV (a Monza that he had for twenty-two years) after seeing an article in late 1976 that cast doubts on the survival of the marque: 'I knew then that I had to buy one, now or never.'

The author (left) with Dave Kay in his West Midlands workshop, spring 1999.

Kay's love for MVs stems from the days when Gilera and MV battled on the track, inspired by the exploits of such men as McIntyre, Surtees, Duke, Hocking and Hailwood.

In 1979, Dave Kay was a founder member, together with John Safe (first Chairman), Peter Ide, Peter Eacott and Alan Elderton, of the MV Agusta Owners' Club GB.

During 1983, on the occasion of the second World MV Rally in Gallerate, the 750S ridden by Dave and his son Mark (then 14 years old) was stolen outside their hotel. This acted as a catalyst for the Kay Engineering effort, which has seen the manufacture not only of parts, but complete engines and, even later still, of complete motorcycles, including Replica MV and Gilera four-cylinder racers. These replicas are superb examples, and the first complete one, a

500 MV, won the premier award at the 1989 Belle Vue Classic Bike Show.

Most controversial of all was the MV racing sidecar outfit (*see* Chapter 11) that Dave and Mark Kay raced during the 1980s. This brought the two into direct conflict with the CRMC (Classic Racing Motorcycle Club), and, in the main, race-goers were denied the chance to see this beautiful machine. Even so, both Dave and Mark chalked up victories, both having Richard Battison as their passenger.

Dave Kay has talked of his ambition for the future: 'To build a copy of Phil Read's 1974 World Championship 500 MV four racer, with its four-valves-per-cylinder technology.' Together with George Beale (of Benelli replica fame), Kay has done his bit in re-creating the finest motorcycles of yesteryear, and, unlike many others, are honest enough to call their bikes replicas!

of the motorcycle sporting scene on a world-wide basis.

In years gone by, when a racer became outdated, it was simply put to one side. Since the beginning of the 1980s, this trend has been

reversed, and thousands of previously unwanted machines have been put back in service. At first, it was simply the original bikes being re-raced, often by their original riders, but, with the rises in values towards

(Above) *This Dave Kay-built MV special 500cc four won 'Best in Show' at the 1989 Classic Bike Show in Belle Vue, Manchester.*

Close-up of the Kay 500cc MV chain-drive gearbox.

the end of the 1980s, many bikes started to change hands for super-inflated prices. They became too valuable to risk on the track.

Except at certain large meetings, four-cylinder MVs have never been very welcome to the ranks of classic racing. However,

The Kay Ferrari

(Above) *The only one of its kind in the world, the Kay-built Ferrari four-cylinder; pure art on two wheels.*

Dave Kay during construction of the Ferrari four-cylinder, c. 1996. An excellent view of the frame and engine assemblies.

The first complete Kay MV special was the 500SS that won the Best-in-Show award at the Classic Bike Show in April 1989. This led directly to the construction of a new, larger engine unit, completed in June 1990. Displacing 901cc (70×62mm), this unit was to form the basis for a one-only motorcycle bearing the world-famous 'prancing horse' Ferrari logo. To avoid copyright problems, Dave Kay's friend and fellow MV enthusiast Rodney Timson wrote to Ferrari. On 23 May 1990, he received the following reply from Piero Ferrari himself, son of the founder Enzo Ferrari:

I received your letter dated 14 May 1990 and give you my approval for you to place the Ferrari badge on your motorcycle. Good luck to you and thank you for your interest.

The next stage was the construction of a frame. Unlike the MV effort, Kay's Ferrari featured a frame that used the engine as a stressed member; this was built by Denny Barber (a former British grass-track champion) from Attleborough, Norfolk. The design also allowed the engine to be removed from the motorcycle by simply taking out eight bolts. Kay went for a pair of Astralite spun-aluminium riveted wheels, which he thought were in the Ferrari tradition.

It was to take two years before the definitive shape for the bodywork was decided. This was made from hand-beaten aluminium by Shropshire craftsman Terry Hall. The styling, particularly for the side panels and the seat base, was inspired by two of Ferrari's best-known cars, the Testarossa and the 280GTO (Dave Kay's favourite four-wheeler).

Detail items took up much time and added greatly to the overall cost of the project. For a start, there were the one-off cam covers with their prancing horse logos. Demon Tweaks (a few miles away in Tamworth) were to supply the digital instrumentation, while the console for these was milled from solid aluminium by Dave Kay's son, Mark.

The genuine Ferrari fuel-filler cap cost a massive £485 from London Ferrari agents Marenello, while the enamel badges for the fuel tank cost another £520 for the pair, making the cost for the fuel tank alone a staggering £2,200, including manufacture and painting!

Finally, in early 1995, the Ferrari project bike was finished, making its public debut at the Stafford Classic Bike Show in April. At the prestigious Festival of Speed at Silverstone, it was put in a special garden thanks to the involvement of Lord Hesketh of the BRDC (British Racing Drivers Club).

The Ferrari cockpit.

Top end of the Kay-Ferrari engine, showing both the cylinder head and carburettors.

Dave Kay had planned on keeping the machine himself, but in the end he sold it (to a very wealthy buyer) to raise cash for his next project – a 500 Grand Prix Gilera four-cylinder replica. But that's another story …

135

together with the six-cylinder Honda 250/297cc ex-works machines, the various multi-cylinder MVs are without doubt the most popular bikes with spectators. Their magical sound is unmatched by anything else on the track, and the stars of yesteryear who ride them, like Surtees and Agostini, are a great draw.

(Above) *MV Club day, Mallory Park, 16 June 1992. Track days are an integral part of the Classic scene, allowing owners to experience a race circuit, without entering a meeting.*

MV enthusiast Peter Jones (350 MV four, 49) at the Italian Owners' Club weekend at Cadwell Park, 25/26 July 1992.

PARADES

Parading is a way of sampling the race action without actually racing. At many classic race meetings, including the CRMC (Classic Racing Motorcycle Club), a number of parades are run alongside the race programme. To take part in the parades, riders are required to enter the meeting, pay the entry fee, go through the pre-race scrutineering process and park alongside the real racers, but no competition licence is required. It is ideal for those whose racing days are over, for whatever reason, or who want to try their bike out without actually competing in a race. They will be allocated a number, but under CRMC rules they must display a small 'P' on each number background.

A variety of different bikes usually participate in parades, from carefully prepared stock sports roadsters to full Grand Prix machines. Unlike pukka racing, bikes from many capacity classes can be circulating at the same time, although riders are usually

Albert Bold

In the history of MV people, American Albert Bold ranks alongside Dave Kay. Both men eat, breathe and sleep MV Agusta fours. Both love the marque, and build their own machines, based around the classic air-cooled dohc four, but improved by modern engineering practices and a vast array of expertise built up over the years. Both Bold and Kay have received world-wide publicity through a series of magazine articles on their machines, and via their personal involvement based on the aura of the four-cylinder MV.

Journalist Alan Cathcart has described Bold, who is based in Zieglerville, Pennsylvania, as follows:

> I wouldn't exactly call Albert Bold eccentric, more just plain crazy: how else to explain that Messianic light shining in those deep, blue eyes, proof of his abiding love for all things red and loud and made in Italy? John-Boy Walton lookalike, ten years on and thin on top, Albert's craziness manifests itself in all sorts of ways, mostly to do with two wheels. Not many people use the bike they go vintage racing with to commute to work on daily, still less when it's a tricked-out four-cylinder MV Agusta special with open meggas and a first gear high enough to beat America's 'double-nickel' 55mph speed limit without changing up. Crazy Albert does. Still fewer people are such committed street squirrels that they'd rather ride everywhere at full tilt even in a land infested by radar. Crazy Albert does. Still fewer people are skilled enough machinists to scorn the idea of buying hardware like brake discs or exhaust pipes off the shelf, preferring to make them themselves. Fewest of all would take their title winning MV vintage racer, buy themselves a GSX-R1100 for day-to-day street squirreling, and tear into the MV to make it lighter, faster, neater and better. Crazy Albert did.

Over the last few years, Albert Bold has built a series of hot MV specials, all of which show his skill as an engineer, and his painstaking dedication to the art of weight reduction. In his own words, Bold spends more than two thousand hours building a bike, and his MVs have been not only for racing, but also for fast road use. Albert Bold is not crazy, but a dedicated genius for whom only perfection is good enough.

(Above) Albert Bold's first MV four race bike took some 2,000 man hours to build, and went on to win the American Classic championship title. Pictured around 1987.

Bold-MV street bike: 862cc MV power unit, four 32mm Malossi-Dell'Orto carbs and Magni exhaust.

A Dream Fulfilled

Pat Sefton (33) guns the MV away from the start of the 1992 Senior Manx Grand Prix. Number 34 is Paul Marks (750 GSX-R Suzuki).

Twenty years after the last works MV had been ridden (by Giacomo Agostini) in the Isle of Man TT in 1972, an MV returned to do battle over the fearsome Mountain Circuit in the 1992 Manx Grand Prix. The rider was Pat Sefton, the bike was a race-prepared 750 roadster with chain final drive, and the driving force behind the venture was MV super-enthusiast Dave Kay.

The dream to race the MV came true following earlier racing exploits (*see* Chapter 11), which had seen both Dave and his son Mark race an MV on three wheels. Mark had then used the same engine to win the 1988 Classic sidecar race over the Southern 100 course near Castletown, during TT week.

The dream was to race a solo MV four in the Isle of Man – ultimately over the full Mountain Circuit. The Kays' ambition was only partly fulfilled in 1990, when Pat Sefton had taken part in the 1300cc Classic TT in late May 1990 over the Southern 100 Castletown course. When lying fourth on the last lap, Sefton had crashed out in wet conditions after clipping a straw bale with his gear-change lever.

Friday, 4 September 1992 was the date for the 750cc Senior Manx Grand Prix. After completing fourteen laps during the practice week, the Kay-prepared MV, ridden by Pat Sefton, lined up on the Glencruthery Road at 1pm. In a wet and windy race –

and competing against modern machinery (no classics this time) – the MV came home in fortieth position out of 105 starters, in the 4-lap 131-mile (210-km) race, with a race average speed of over 95mph (150km/h). The MV's fastest lap, during practice, was made in 22 minutes 30.6 seconds, an average speed of 100.57mph (160.9km/h). Among scores of Hondas, Kawasakis, Suzukis and Yamahas, the only other non-Japanese machines in the entry, besides the MV, was a 600 Rotax single and a 750 Norton twin. Their lap speeds were way down on the four-cylinder MV's.

The result was immaterial; the sight and sound of a four-cylinder MV in full flight around the TT course was enough. As Dave Kay related afterwards, 'The tears were streaming down my face in the grandstand. I'd waited ten years to see and hear this!'

In wet and windy conditions, Pat Sefton completed the 1992 Senior Manx Grand Prix at an average speed of 95mph (152km/h), finishing fortieth out of 105 starters – a magnificent performance considering the opposition was almost all modern Japanese superbikes.

Team Obsolete

In its August 1993 issue, *Classic Bike* reported 'New York's Team Obsolete is the apparent winner of a bitter six-year legal battle over ownership of legendary MV Agusta racers.' The story went on as follows:

> A recent settlement of several lawsuits in New York and in La Spezia, Italy, where the collection of Grand Prix machines has been stored during the dispute, has confirmed that 17 Grand Prix racers and a roomful of spares are the property of TO [Team Obsolete], run by Brooklyn lawyer Robert Iannucci.
>
> According to Iannucci, 'They're mine, and we're going to race some – as Team Obsolete always intended to, since buying them in 1986.'

The complete listing of former works MVs awarded to Iannucci was as follows:

5 1970s 500cc fours
2 1970s 350cc fours
1 1970s 750cc four
2 1960s 500cc threes
1 1960s 350cc three
1 1960s 500cc four-valve four
2 1970s 500cc Boxer flat fours
1 1950s 125cc dohc single
1 1950s 125cc four-valve twin
1 1950s 203cc dohc single

At the time, it was reported that some of the machines, including the two 500cc Boxer flat four prototypes, which had never been raced, were being sold off. Iannucci's former partner in Team Obsolete, Jeff Elghanayan, who had strenuously contested ownership of Team Obsolete bikes, was also awarded custody of three other machines.

The whole legal saga had begun back in 1988 with a court case in New York. Elghanayan is believed to have instigated several other lawsuits in both the USA and Italy in attempts to claim ownership of the MV purchase, when the two men fell out in 1987.

The entire purchase was kept during the period of dispute at

Part of the haul of MV racers which Team Obsolete purchased from Gruppo Agusta in 1986; the subject of a bitter six-year legal wrangle.

the racing headquarters in La Spezia of Roberto Gallina (former racer and, later, Suzuki GP team manager), following its purchase from Agusta by Team Obsolete in 1986.

Iannucci was reluctant to reveal details of the protracted ownership wrangle, but *Classic Bike* estimated it to have cost 'over $1 million in legal fees alone'.

A judge in La Spezia consulted many informed sources, including Giacomo Agostini, former MV mechanic Ruggero Mazza and Italian magazine *Motociclismo*'s editor and director.

Within days of the settlement, Iannucci took thirteen of the MVs to Italy's Misano Historic Grand Prix circuit festival, where TO rider John Cronshaw rode 350 and 500cc triples in front of delighted fans. However, the whole sorry saga of this hotly disputed legal wrangle should act as a reminder that money really can be the root of all evil.

Two of the Team Obsolete collection: a 350 four (left) and a 500 four (right), both final versions. The 'AGV', 'Castrol' and 'Gruppo Agusta' logos on the fairing denote the original Team Obsolete MV sponsors.

despatched from the starting grid in small groups. Sessions start and finish in the same way as 'real' races.

Those taking part in the parades will also have the same opportunity as actual racers for at least one pre-parade practice session. Even though they will not be racing for real, attention to preparation, machine condition, riding gear, tyres, and the like, are still vitally important.

(Above) *Team Obsolete machines, Mallory Park Post TT, June 1997. Number 21 is one of the last 350 fours built. The other two machines are three-cylinder models. Number 7 in the foreground is George Beale's Benelli four replica.*

Kay Engineering one-off MV special using a magnesium 800cc engine with Dell'Orto 40 DCOE twin-choke carbs, 1998.

Magni Chain-Drive Conversion

Produced by Elaborazioni Magni de Magni Giovanni of Via Milano 56, 21017 Samarate (VA), Italy, the chain-drive conversion kit is not only one of the most important after-market accessories for the 1965–78 MV four-cylinder roadster with shaft final drive, but also one of the most expensive. However, this would have to be the one change to make – the standard set-up is both heavy and not without fault.

The chain-drive kit conversion comprises the following components:

- box-section swinging arm and chainguard; cush-drive/sprocket carrier and sprocket, which can be bolted to the standard wire wheel; chain (530 size);
- axle for swinging arm and a pair of threaded bushes – the latter screw into the standard threaded frame lugs;
- replacement dyna-start support pin; replacement gearbox casting with new fifth drive gear, gear selector (the splined shaft of which is shorter than standard), all bearings, a GRP chain guide, replacement crossover shaft for the gear change, gearbox sprocket, and an operating lever for the clutch and its support;
- pieces for frame modification – two tube sections and a threaded bush. The latter is welded into the cross-brace above the swinging-arm axis (a bolt is used to eliminate engine movement that would otherwise be caused by the pull of the chain – the end of the bolt rests on the rear edge of the replacement gearbox casting cover).

Three instruction sheets have been provided with some kits. However, according to Dave Kay, 'Most people appear to ignore these, instead use common sense and basic engineering practices.'

It may be discovered on receipt of the kit that a replacement splined lever for the gear-change linkage has been omitted, and that the splining of the original lever will not match the replacement crossover shaft. Use a lever off a post-1979 Ducati bevel V-twin (with Darmah-type bottom end), suitably shortened and drilled.

Arturo Magni (right) and John Surtees, two great men who helped shape the four-cylinder MV Agusta.

To allow sufficient chain clearance along the top run, the existing battery carrier must be shortened and a smaller (narrower) 28-amp battery substituted. If this is not done, sooner or later the chain will cut through the base of the battery.

Before actually welding in the replacement sections, it is advised that the crankcase, centre stand and dyna-start are temporarily fitted into the frame – remembering to check first for clearance.

The replacement cross-brace will also need to be positioned so that the standard centre can be retained. Also check that the chain does not rub. If it does, a piece of neoprene rubber can be fitted to the top of (or around) the offending tube.

Finally, use the TIG process for welding on the replacement sections.

Magni logo. Besides his well-known MV connections, Arturo Magni has also more recently been involved with a series of specials powered by BMW flat-twin and Moto Guzzi V-twin engines.

11   Three Wheels

THE SWISS CONNECTION

The first attempt to build a racing sidecar powered by a four-cylinder MV Agusta power unit came in the late 1960s, when two Swiss 'chair' specialists – Edgar Strub and Hans-Peter Hubacher – sought to transform one of the recently released 600 four roadsters into a competitive outfit. Making its public debut in early 1968, their racing sidecar showed considerable engineering skill, but in practice the finished device never fully replaced Hubacher's twin-cylinder BMW Rennsport outfit. The Strub-Hubacher MV took many hours of hard work to make the switch from road-going touring solo to a road-racing three-wheeler. At the end there was very little remaining of the MV, except for the engine assembly and shaft final drive. In the process, the motor was sleeved down to 500cc, while the cylinder head was reworked to include four carburettors (on the street, the 600 tourer had only two). These instruments were 27mm Dell'Orto SS1s – the same as on the factory racing fours of the period.

Ultimately, Strub and Hubacher were forced to accept that the down-sized 600cc

(Opposite page) The Kay MV sidecar was not the first of its kind, but it was certainly the most controversial, sparking objections from within the CRMC (Classic Racing Motorcycle Club). Was it right to stop the public seeing this fabulous machine in action?

touring engine with its shaft final drive was not going to beat their existing BMW-engined creation.

THE KAY THREE-WHEELER

At the centre of controversy from the moment it was unveiled in 1985, the Kay MV four-cylinder racing sidecar was a neat piece of kit. Maybe it was too good, getting its opponents up in arms before the flag had dropped.

Dave Kay, a founder member of the MV Agusta Owners' Club in the early 1980s, was also a highly skilled engineer and a sidecar racer of some repute. His career on three wheels began as a passenger to friend and fellow MV enthusiast Bill Johnson. At first, the pair campaigned a very quick Laverda Jota-powered outfit at club meetings during the late 1970s, and then Kay became a driver at the CRMC (Classic Racing Motorcycle Club) events during the early 1980s with an immaculate 1200 NSU. This involvement with sidecars, coupled with a life-long love affair with four-cylinder MV Agustas, was bound to coincide sooner or later. For years, he had discussed putting an MV four engine in an outfit for classic racing, and by the middle of the 1980s he had done it.

The result was a superbly engineered machine that was a credit not only to Dave Kay, but also to Ron Graham, his friend and business associate, who had contributed both money and expertise to the project.

143

The Kay outfit consisted of a 1972 750 Sport engine, housed in a chassis purpose built by John Derbyshire. The design was such that the power unit, which was held in place by seven mounting bolts, could be completely stripped from the chassis in under ten minutes – easy access being all-important in a racing machine. If only the top section of the engine needed attention, the cylinder head and barrels could be removed with the engine *in situ*. The near-side frame tubes were removable for access to the clutch and for sprocket changing (a Magni chain drive was part of the specification) on the crossover shaft.

The massive box-section swinging arm was controlled by a single Dutch Koni shock absorber. This was mounted on eccentric bushes, which enabled the drive chain to be adjusted in less than a minute. Car-type cast-alloy wheels were used, running on 145x10 Dunlop Green Spot tyres. Braking was by Lockheed. The system worked with the handlebar lever operating one of the twin front discs, while the foot-operated brake pedal operated *two* master cylinders

(Above) *The original machine; but it was the Kay MV sidecar 'Mark 2' that was actually raced.*

The engine of the Kay MV sidecar racer was based on a 750S from the early 1970s.

– the first the other front disc, and the second to work the rear wheel and sidecar wheel discs. There were four discs in all, with twin-piston calipers on each. All wheels were quickly detachable, leaving the calipers and discs in position.

The 750 Sport engine was largely of standard specification, but blueprinted to achieve the best possible power output, while retaining a high level of reliability. Several road-going components had, however, been removed, most notably the Dynostart, which alone saved 20lb (9kg). Although the standard five-speed gearbox was retained, a chain final-drive conversion by Arturo Magni, one-time MV racing team manager, had been fitted. Not only did this absorb less power than the original shaft arrangement, but, perhaps

Impressive engine-top view of the Kay MV sidecar, showing cylinder head, four carburettors and a plate dedicated to the memory of works racer Les Graham and Club member Dave Ambler.

Magni chain-drive conversion, a feature of the Kay MV machine.

Front-end details of the Kay MV sidecar – massive forks with Dutch Koni shock absorbers, hydraulic steering damper and Lockheed two-piston brake caliper.

more importantly, it also allowed the gearing to be altered quickly for circuit conditions.

In view of the extra stress expected, an additional support bearing had been machined into the outside spyder outrigger to give extra support to the fifth gear on the mainshaft.

A higher-lift camshaft profile was employed, which, together with an increase in compression ratio from 9.5:1 to 10:1, and modified inlet tracts to suit the 30mm Dell'Orto PHF pumper carbs, pushed maximum power to 90bhp. The effective range of power spanned 5,000 to 10,000rpm. The special one-off four-pipe exhaust system was

the work of Midlands engineer Dave Kerby, to Dave Kay's own specification.

Ignition was by coil and battery via a car-type distributor (ex-Fiat). The battery was housed at the front of the chair under the perspex bubble, with the fuel pump under the rear of the sidecar platform, hidden away neatly behind a skirt together with the fuel filter and pressure regulator.

The MV engine is peculiar because the final drive is close to the centreline of the motor. To achieve a proper relationship between front and rear wheel it was necessary to fit a crossover shaft to drive the rear wheel on the nearside – standard sidecar practice on Imp-powered specials. The shaft used was a Z400 Kawasaki mainshaft on which were specially made sprockets. The centres of these sprockets were cut by Ron Graham using a Spark Eroder. Although cutting with one of these is fully automatic, it took five hours to make up the required tool; as each electrode is only good for cutting six sprockets, Ron missed out on some drinking time for a week or so! As both crossover shaft and engine mainshaft are of fixed centres, it was necessary to install an eccentrically mounted nylon jockey pulley to take up slack in the external primary drive chain.

Originally fitted with a 1960 DMD dustbin roadster fairing (but suitably modified with airscoops and quickly detachable mountings), the Kay MV racing sidecar outfit was fitted with a small handlebar fairing.

An aluminium fuel tank was hidden from view beneath the sidecar wheel arch; three holes in the wheel fairing were to direct cool air to the sidecar wheel disc.

Following some forty laps of the Cadwell Park club circuit at a practice day, the Kay MV was entered for a Retford Club meeting at the same venue. Although competing against modern machinery, the bike was not disgraced, despite a fourteen-year old engine and a chassis constructed to an out-of-date

Testing the Kay MV sidecar at Cadwell Park in 1985. Note the fairing – a 1960 DMD dustbin type – from a road-going solo!

design built specifically with classic racing in mind.

However, this was where the problems started. Soon after completion, the Kay outfit was banned from CRMC classic racing events, after wrangles over its eligibility.

ISLE OF MAN VICTORY

MV Agusta had won just about every solo class in the Isle of Man, but not the sidecar class. However, all this changed in 1988 when the Classic TT was moved to the bumpy Billown circuit near Castletown (the home of the annual Southern 100 races), and classic sidecars were included for the first time. This time, the driver was Dave Kay's 20-year-old son Mark, passengered by Richard Battison.

Commenting later, Mark Kay said, 'We geared for 8,500 maximum revs because of the bumps, we normally use 9,500 on short circuits, but over there they have so many bumps that you could be up to 10,000 and more before you could shut it off. So we just

geared it high and used the torque of the motor.'

After bad weather during practice, race day, 6 June, dawned dry and sunny. By the second lap, Kay junior and Battison were up to third place; on the third lap, the pairing made their first challenge for the lead, but spun out on the notorious Four Ways Hairpin Bend. After this setback, it took another lap to get back with the leaders again, and this time the MV got past, and stayed there, increasing its lead on each lap. At the finish, it was thirteen seconds ahead of its nearest rival.

The win was an unforgettable first, not only for the victors, but also for Dave Kay, who had worked so hard in preparing the machine. It was the first time in over a decade that an MV had won a TT race, so the team was in great demand among both radio and newspaper reporters. The cup won by Mark Kay and Richard Battison was given to Mark's dad 'for all his invaluable help and advice over the years, and considering the lack of any real trade sponsorship'. 'I cannot stress too highly the debt of gratitude we owe him,' said Mark after the victory.

Sadly, the backbiting that had dogged the outfit throughout its life returned with a swingeing letter in *Motor Cycle News* saying that the MV outfit was not a classic. The same criticism was also to appear in the CRMC's newsletter *Open Megga*.

The Kay MV winning the 1988 Classic Sidecar TT over the Billown Circuit near Castletown, crewed by driver Mark Kay (then 20 years old) and passengered by Richard Battison.

(Below) *Mark Kay and Richard Battison rounding Mallory Park's Hairpin, 7 June 1992.*

12 Three and Six Cylinders

500 SIX

The six-cylinder came before the three in MV's grand strategy; in other words, in Count Domenico Agusta's overall racing plan. In fact, the six, as a 500 blue riband Grand Prix engine, was created in response to rival Gilera's mighty four-cylinder, and also in the light of Moto Guzzi's V8 model.

By the mid-1950s, the development of the Italian Grand Prix motorcycle was at its peak, spurred on by intense rivalry, never before seen and unlikely to be repeated again, between the big guns of the Italian bike industry. Even though Count Domenico had signed the talented John Surtees to race his company's big-bore machines, even the young Englishman's skills, it was reasoned, would be hard pressed to defeat the best that Gilera and Moto Guzzi could throw at MV. It was decided that the Gallerate concern needed its own multi-cylinder flagship – in the shape of a newly created six-cylinder powerhouse. Much of the early testing was done by rider/team manager Nello Pagani (father of racing son, Alberto).

Unfortunately, the 499.2cc (48 × 46mm) dohc across-the-frame *sei* (six) was never to fulfil its creator's dreams. Wheeled out for the first time in public at the Monza autodrome during practice for the 1957 Grand Prix of Nations, its arrival coincided with the

Monza, August 1957. MV's team manager Nello Pagani about to test the new 500 six-cylinder model. The withdrawal of Gilera and Moto Guzzi at the end of that year meant that it was never needed. This rare photograph is one of only a few in existence – and probably the only one showing the original full 'dustbin' streamlining that was banned by the FIM shortly afterwards.

tripartite pact between Gilera, Moto Guzzi and FB Mondial to withdraw from Grand Prix racing at the end of the 1957 season. Effectively, this meant that the Monza meeting was the final round of the 1957 championship series.

The effect of the withdrawal of Gilera and Guzzi (Mondial only contested the lightweight categories) was to render the new MV six an almost redundant exercise. MV's

existing four-cylinder 500 was more than capable of dealing with the opposition that was left: privately entered British singles, plus the odd pushrod twin-cylinder such as the Matchless G45 and race-kitted Triumphs.

All this was a great pity and it deprived the race-going public of a chance to see the new MV six do battle with Guzzi's V8 – when both were fully developed, which of course never happened.

Another view of the fabulous 499.2cc (48 × 46mm) six, at Monza on 24 February 1958, showing the dolphin fairing, which was new for that year.

The only known photograph showing the new 500 six-cylinder model on the day it was raced by John Hartle (its only outing) at the Italian Grand Prix in September 1958. It was destined to retire at half distance in the event.

500 Six (1957)

Engine

Type	Air-cooled dohc six, across-the-frame
Bore and stroke	48 × 46mm
Capacity	499.2cc
Compression ratio	10.8:1
Carburation	Six Dell'Orto 26mm SS1 26A carburettors with remotely mounted float chambers
Lubrication	Wet sump
Max. power (at crank)	75bhp @ 15,000rpm
Fuel tank capacity	4.84 gallons (22 litres)

Transmission

Gearbox	Six speeds
Clutch	Wet, multi-plate
Primary drive	Gear
Final drive	Chain
Ignition	Magneto

Frame

Tubular steel, closed duplex cradle

Suspension and steering

Suspension	front	Teledraulic fork
	rear	Swinging arm with hydraulic shock absorbers
Tyres	front	3.00 × 18
	rear	3.50 × 18

Brakes

	front	Full-width drums, 4LS 260mm
	rear	Full-width drums, 2LS 190mm

Dimensions

Dry weight	320lb (145kg)

Performance

Top speed	149mph (240km/h)

The reason for moving from four to six cylinders was simple – with the higher engine revolutions, the engineering team expected to achieve higher rev limits and, therefore, better performance.

The layout of the MV six followed that of the four: the cylinder block was set across-the-frame and inclined forward at 10 degrees from vertical. Primary drive was via a gear train driven off the end of the crankshaft, but the gearbox had six instead of five ratios.

Another innovation over the four was the use of chrome-bore aluminium cylinders. The MV *sei* featured an open-frame structure at the rear, with a detachable lower cradle (for ease of engine removal), which was clearly modelled on its four-cylinder brother, and many of the cycle parts came from the same source.

Other details of the machine's specifications included six 24mm Dell'Orto SS1A carbs, magneto ignition, teledraulic front fork, 18in wheels and a 22-litre fuel tank. Dry weight was 320lb (145kg). The engine produced a claimed 75bhp at 15,000rpm.

On its public debut at Monza in September 1957, the machine was equipped with a fully enclosed dustbin fairing. The FIM banned these at the end of that year and subsequent appearances saw the six clothed in a conventional dolphin-type fairing.

Although it was used several times during practice, the machine's only race came a year later at Monza during the Italian GP in September 1958. Here, ridden by John Hartle, it proved fast, but temperamental. Hartle was in fact the last rider to get away from the start line, the six being reluctant to fire up; however, once under way, the six gobbled up the opposition to storm from 26th on the first lap to 4th, when the engine cut out suddenly just after half distance. The 35-lap, 125.5-mile race was won by MV team leader John Surtees, in 1 hour 5 minutes 31.4 seconds, at an average speed of 114.51mph (183.2km/h).

Surtees also set the fastest lap at 115.98mph (185.5km/h). It is worth noting that, besides Hartle and Surtees, Remo Venturi (2nd), Umberto Masetti (3rd), Carlo Bandirola (5th) and Giuseppe Cantoni (8th) were also mounted on MV four-cylinder models.

The six was then mothballed. It was a great pity as, with its exit, 500cc race development effectively became stalled for almost a decade, before a larger version of the three-cylinder 350 was constructed towards the end of the 1960s.

350 SIX

Another era and another generation saw a new six appear during 1969 but, again, events were to conspire against it, almost from the start. This time, it was not rival manufacturers that posed the problem, but the sports governing body, the FIM. It decreed that, from the 1970 season, motorcycles would be restricted to only six speeds, and that 350 and 500 engines could only have a maximum four cylinders; 250s would be restricted to two cylinders, and 50 and 125cc models to only one cylinder.

As MV's new model had six cylinders *and* a seven-speed gearbox, its future was sealed. The public was once again denied the chance of seeing a six-cylinder MV in action, but Agusta was largely to blame as effectively it had missed the boat. Honda had already been racing similar six-cylinder models several years before in both 248 and 297cc engine sizes.

Although, in many ways, the engine design followed the earlier 500 six, the 1969 350 was considerably different in one important area – chassis design. The smaller six had a duplex full-cradle frame, instead of one with detachable bottom rails, and the swinging arm was in square tubing, not rounded as on the original. Also, the cycle

parts were largely from the three, rather than the four, as on the old 500 model.

The 350 did share one feature with its bigger brother – the primary drive was taken from one end of the crankshaft. Originally, the 348.8cc (43.5×39.5mm) across-the-frame six featured the largely experimental electronic ignition system. However, although the theory was good, the technology at that time was not fully up and running, and battery/coil with contact breakers was soon adopted.

Even from the earliest bench tests, the engine had been found to produce over 70bhp – an excellent figure for the period. Ultimately, MV was to claim 72bhp at 16,000rpm and 155.5mph (250km/h). This was all rather academic, as the six was not allowed to be raced, but MV did take the model to several GPs for testing during 1969 (and occasionally afterwards).

350 THREE

The reasoning behind the three-cylinder MV was based on the introduction in 1962 of Honda's larger-engined version of its 250cc four, the 285cc model, for the 350cc class. Soon, a full-sized 350cc Honda four was introduced. Flown to Europe, this proved so superior to the ageing 350 MV four (*see* Chapter 4) that Honda's team leader simply outsped MV's Mike Hailwood in the remaining GPs of that year, and won the world title. In typical fashion, Count Domenico Agusta decided to retire from the class rather than get soundly beaten (as he had already done when the 250cc Honda four trounced the MV twin in 1961). This only left the 500cc class, in which Hailwood took the title in 1963, 1964 and 1965.

In that final year, an entirely new combination made its debut into the Grand Prix arena – Giacomo Agostini on the three-cylinder MV Agusta. This was just the tonic for the autocratic Count, whose dream had always been of an Italian rider on an Italian bike in the larger (350 and 500cc) racing classes.

'Ago' made his name racing for the tiny Moto Morini factory in the 250cc Senior Italian Championships. During 1964, he convincingly beat his former teacher Tarquinio

Provini, who had left Morini to join Benelli at the end of 1963 to ride the four-cylinder Pesaro-built machine, thinking that this would give him an edge over the single-cylinder Morini.

In Ago, Count Domineco Agusta saw someone capable of achieving what he had previously thought an impossible dream, so he sanctioned the building of the three-cylinder model that had first been mooted as long ago as 1958.

As on the six-cylinder models, the 343.9cc (48 × 46mm) across-the-frame triple had its cylinders inclined forward some 10 degrees from the vertical. Other items of the machine's impressive technical specification included a seven-speed transmission, 18in wire wheels with Borrani alloy rims, a 3.5-gallon (16-litre) fuel tank, and a quad-cam 240mm drum front brake of immense size and power. With over 62bhp at 13,500rpm available, the new MV could at last challenge the Japanese dominance of the class. With a maximum speed of 150mph (240km/h), it was not only as rapid as its larger four-cylinder 500 brother, but because of its lighter weight

(Above) The three-cylinder 350 made its debut in 1965. New signing Giacomo Agostini in winning form during the Italian GP that year. He won the race in 51 minutes, 12.5 seconds.

Seen in the MV pit at the Italian GP on Sunday 8 September 1965, the three-cylinder 350 had four valves per cylinder, displaced 349.2cc (55 × 49mm) and revved to over 12,000. The gearbox was a seven-speeder, while maximum power was around 58bhp.

and superior power-to-weight ratio, it was a superior racing motorcycle.

Agusta's new signing, together with the triple, made a sensational Grand Prix debut. The *Motor Cycling* race report of the West German Grand Prix, dated 1 May 1965, said it all: 'Agostini shatters Redman'. The story went on to recount: 'Undefeated throughout

last season's 350cc classic road races, Honda team leader Jim Redman met his match in Giacomo Agostini and the new three-cylinder MVs in Saturday's event.' After a 17-lap (81.77-mile/131-km) wheel-to-wheel duel with the young Italian on the damp, slippery southern loop of the Nürburgring, Redman and the Honda went 'down the road'. The full

---

350 Three (1965)

Engine

Type	Air-cooled dohc three, across-the-frame
Bore and stroke	48 × 46mm
Capacity	343.9cc
Compression ratio	11:1
Carburation	Three Dell'Orto 28mm SS 28B carburettors with remotely mounted float chambers
Lubrication	Wet sump
Max. power (at crank)	62.5bhp @ 13,500rpm
Fuel tank capacity	3.5 gallons (16 litres)

Transmission

Gearbox	Seven speeds
Clutch	Dry, multi-plate
Primary drive	Gear
Final drive	Chain
Ignition	Battery/coil

Frame
Tubular steel, duplex cradle

Suspension and steering

Suspension	front	Teledraulic fork
	rear	Swinging arm with hydraulic shock absorbers
Tyres	front	3.00 × 18
	rear	3.25 × 18

Brakes

	front	Full-width drums, 4LS 240mm
	rear	Full-width drums, 2LS 230mm

Dimensions

Dry weight	256lb (116kg)

Performance

Top speed	149mph (240km/h)

race distance was 20 laps (96.2 miles/
154km). Ago's average speed was 84.75mph
(135.6km/h), his fastest lap (the quickest of
the event) was 87.86mph (140.5km/h). Incidentally, Agostini lapped the entire field,
bar team-mate Hailwood, who was half a
lap adrift on one of the old four-cylinder
models. Although Hailwood won the 500cc
title on an MV that year, he was clearly
upset about the attention lavished on his
Italian team-mate. When Honda made him
an offer of big money to move, he signed on
the dotted line. Meanwhile, 'Ago' went on to
win his first world crown, the first of many.

500 THREE

Throughout 1966, Agostini and Hailwood
fought out some fantastic battles, the Honda
man winning the 350cc title (on a 297cc six),
with the Italian realizing Count Agusta's
dream by taking the blue riband 500cc crown.
MV's success came with an enlarged triple,
rather than an updated four-cylinder model.
At first, the displacement was increased to
377cc (by boring out the cylinders to 55mm),

*The duels between Honda-mounted Mike
Hailwood and Ago's three-cylinder MV in the
Isle of Man during the late 1960s are legendary.*

then 420cc and finally in 1967, 497.9cc (62 ×
55.3mm).

Hailwood and Agostini duplicated their
championship results in 1967, but the MV
rider's success was achieved only after a
titanic struggle with the Englishman, each

*Agostini pilots the new 500
three-cylinder during the 1966
Czech GP. Although he finished
second behind Hailwood, the
Italian went on to become
500cc World Champion, his
first of many as an MV rider.
The 491.2cc (60.5 × 57mm)
triple followed the design of the
smaller model, with its gear-
driven dohc and four valves
per cylinder; maximum power
was 80bhp at 12,000rpm.*

Ago during the 1967 Senior TT. He is shown here at Brandish some 35 miles out. On his larger triple he shattered the lap record from a standing start, only to have the new record beaten by Hailwood (Honda) a lap later. The MV rider was to retire with only one lap to go while leading Hailwood by two seconds.

Giacomo Agostini cresting the Mountain at Cadwell Park, September 1970.

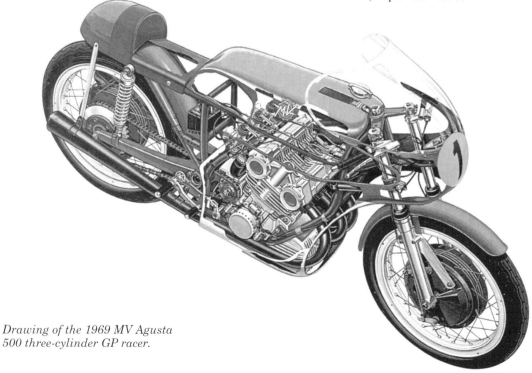

Drawing of the 1969 MV Agusta 500 three-cylinder GP racer.

Frontal view of the 500 three-cylinder engine.

(Right) *The three Dell'Orto SS1 28 A carbs with their pair of float chambers as used on both the 350 and 500cc MV triples.*

SSI 28 A

campione del mondo

350-500cc. m.v. agusta 3c.

(Below) *Three-cylinder crankshaft, pictured in the Team Obsolete workshop during 1995.*

(Above) *The final Grand Prix of 1970, in Barcelona, Spain, over the weekend of 26/27 September. With both the 350 and 500cc world titles already in the bag, Giacomo Agostini did not take part. Angelo Bergamonti was given the responsibility, and responded by winning both races on three-cylinder MVs; this is the 500cc.*

of them winning five of the ten Grands Prix contested. An extra second place made all the difference to Agostini. At Monza, the 500cc race should have brought Hailwood the title, but, two laps from the finish and with a half-lap lead over the Italian, his Honda jammed in gear.

Hailwood's retirement and Honda's withdrawal from the sport made Agostini's task much easier in 1968. He won all seventeen championship races contested that year – all on three-cylinder MVs. Even though the triples were later superseded by new versions of the four-cylinder theme, it was the MV triples that did much to cement Agostini's career as the world's most successful road racer – a position he holds to this day because of the number of championship titles won.

Angelo Bergamonti winning the 350cc race in Barcelona, 1970, on the three-cylinder MV.

13 Cagiva

The run-down Varese production facilities, purchased by Cagiva in summer 1978 after Harley-Davidson had closed its Italian operation.

THE CASTIGLIONI FAMILY

Two families have controlled MV: first, the Agustas and, today, the Castiglionis. The first the motorcycling world heard of the Castiglioni family was during the late 1970s, initially through a modified Suzuki RG500, ridden by future world champion Marco Lucchinelli, and, from September 1978, via a new marque called Cagiva. The Cagiva name was an amalgam, made up from 'CA' for Castiglioni, 'GI' for Giovanni (father of the present owners, Claudio and Gianfranco Castiglioni) and 'VA' for Varese (the family's home town, where its factory was based).

The newly created company was initially headed by president Battisto Lozio and the Castiglioni brothers Claudio and Gianfranco. Its 'factory' was the former Aermaachi-Harley Davidson lakeside facility at Schrianna on the edge of Varese. It had taken over all the plant, machinery and production rights for the range of ten road and off-road motorcycles formerly sold under the AMF Harley-Davidson brand name, which had single-cylinder engines, ranging from 125 to 350cc. For an interim period of one year, the petrol tanks were adorned with an 'HD Cagiva' logo, but, from then onwards, the Harley connection would vanish. (Incidentally, the company had originally chosen a white elephant for its emblem. Fortunately, at the British dealer launch I was able to explain the significance of this image to factory officials, and it was quickly changed to grey!)

During the first few months of its life, Cagiva built an improved version of existing HD lightweights from the spares part stockholding. This 1980 SST 250 single-cylinder two-stroke is typical.

The Castiglioni brothers were successful businessmen from Varese who had seen their original company, founded by their father, become one of the region's largest employers. They had made their wealth manufacturing locks, belt buckles, clamps and the other bits of metalwork found on luggage and handbags. The Castiglioni metal-pressing operation was so efficient that it could actually beat the foreign competition on both price and quality.

Interviewed shortly after the takeover of the former AMF HD factory in autumn 1978, Gianfranco, the elder of the two brothers, gave a simple statement to explain why they had purchased the factory: 'Because we love motorcycles, of course.' It was true; not only had they sponsored Lucchinelli (and also Franco Bonera) *before* becoming motorcycle manufacturers in their own right, but they also subsequently spent billions of lire in a quest to win the 500cc road-racing World Championship, as well as participating enthusiastically off road in motocross and enduro.

THE SECRETS OF SUCCESS

Where did all the money come from? The Castiglionis did use some of their own wealth, and some came in the form of government grants for creating employment. (Indeed, this source of revenue was to remain an important part of the Cagiva group's growth, until such schemes were effectively outlawed by a scandal that hit the Italian political scene in the early 1990s.) The brothers also invested heavily in people. When Cagiva commenced production, a mere 130 workers had been retained from the original 500 or so Harley-Davidson staff. But, of these, over one-quarter were in the R&D department. The Castiglioni brothers also spared no effort in recruiting the right people, including several technicians formerly employed by the now defunct MV Agusta race team, top Dutch two-stroke engineer and tuner Jan Thiel, and, of course, former staff who had built and tuned the Harley-Davidson machines that had brought Walter Villa no less than four world road-racing championship titles in the mid-1970s.

When Cagiva entered the two-wheel world as a manufacturer, it had no new designs and no image (in fact, hardly anyone could pronounce the name correctly!). So-called informed industry specialists gave the company no chance of success in a highly competitive market on the verge of a world recession.

However, the key was that, although the Castiglionis certainly loved bikes, they were also used to thinking like successful businessmen. By emphasizing the fundamentals – efficient manufacturing, high quality, competitive pricing (by Italian standards of the time) – and authorizing new models, the fledgling company was able to achieve the seemingly impossible task of growing from nothing to be a world power within a decade. No European company had managed to do the same in recent times.

Cagiva's approach would have counted for little without a buoyant home market, however, and in this respect the company was extremely fortunate. At the time of its launch, there was virtually no competition from the Japanese, who were barred from the Italian domestic market in the capacity class on which Cagiva was then concentrating – up to 350cc. Later, the Japanese, notably Honda and Yamaha, would get around the embargoes by building plants in Italy themselves.

The new company also had the added advantage of a considerable profit generated by selling the old designs and spare parts stock gained in the purchase of the former

Harley-Davidson factory. During the first few months, and in the first years of the 1980s, they made a killing. Much of this was down to the success of one model, the SST125, which proved to be the top-selling motorcycle in Italy in the all-important 125cc category during 1979–82.

Another reason for Cagiva's early success was its unbiased approach. Unlike their Italian rivals at the time, they were not too proud to bargain, or to examine in detail all the latest Japanese hardware. Cagiva's engineering team was not only fully aware of the competition, but also given the go-ahead to incorporate the best features into its own creations.

FIRST IN-HOUSE PROJECT

One of the very first projects begun by Cagiva engineers after the September 1978 takeover was the development of a liquid-cooled 125cc motocross bike. The prototype arrived in 1979, and went on sale the following May. (No Japanese factory had a production liquid-cooled off-road racer on the market until 1981.)

Cagiva's first in-house design of the new era was the WMX125 motocrosser. Produced in 1980, it had the distinction of being the world's first water-cooled production dirt bike.

For over a decade, the Castiglioni family spent millions in the quest for Cagiva honours in 500cc GPs. The British Grand Prix, Donnington Park 1982 – Cagiva-mounted Jon Ekerold leading Ron Haslam (Honda) at Becketts.

An enduro version, albeit air-cooled, was the next 'new' Cagiva, in spring 1981. Although the engineering team had no real experience of modern two-stroke dirtbikes, the Cagiva 125 motocross and enduro models (coded WMX and RX respectively) matched anything in the world, being both fast and reliable. In 1981, Dutchman Bart Smith broke the 125cc world speed record in a streamliner powered by a tuned version of the WMX125 motocross engine, with a speed of 257.069km/h (159.62mph).

At the end of 1981, the biannual Milan Show saw a whole host of new Cagiva models, including the new company's first four-stroke, the 350 Ala Rossa, a single-cylinder ohc trail bike. The 500 Grand Prix class had also seen the arrival of a new Yamaha-inspired four-cylinder two-stroke, which made its debut at the West German GP earlier that same year.

EXPANDING THE MARKET

In 1980, Cagiva built 13,000 bikes; by 1982, this figure had increased to 40,000. At the same time, there was an increase in the workforce, of up to 300 employees, of which 50 were R&D staff. In 1981, the first foreign plant opened, when a factory in Venezuela began producing Cagivas for South America from parts made in Italy. Several other overseas projects followed. As early as 1981, talks had also taken place with the Communist authorities in Moscow about the possibility of supplying Cagiva expertise to the USSR in the same way as Fiat had done earlier in the four-wheel world. In the end, the Soviet connection was to lead to Russian riders campaigning an air-cooled 500 Cagiva motocross model in the world series during 1982.

The Castiglioni brothers did not give up easily; if one avenue failed, they simply picked another. Their next project was intended to broaden their motorcycle range upwards. Sales director Luigi Giacometti made a visit to England, to the Daventry works of Hesketh Engineering, which had gone into receivership during 1982. This too was to draw a blank, as Giacometti soon discovered that Hesketh farmed out the majority of its component production to outside

Cagiva 500 four-cylinder two-stroke, as it was in 1983.

suppliers. The Cagiva sales director would comment:

> They [Hesketh] had nothing to offer us. The receiver wanted £150,000 for a pile of drawings and papers.

DUCATI–CAGIVA PRESS CONFERENCE

Cagiva was to find the answer to its search much closer to home – in Bologna, the home of Ducati Meccanica. For many years, although seen as charismatic by the press and public alike, Ducati had been burdened by state control from far-off Rome, and had been consistently unprofitable.

On 2 June 1983, Cagiva and Ducati executives called a joint press conference. Held on neutral territory in Milan, this announced to the world that Ducati had contracted to supply Cagiva with engines – intended for the latter company's new range of larger-capacity motorcycles (from 350 through to 1000cc). It was stated that this agreement would initially run over seven years. The Ducati name was to remain on

Cagiva concluded an agreement with Ducati in June 1983, with the latter's V-twin engine being sold to the former. The first results came with the Alazzura roadster (350 seen here) and Elefant on/off-road model.

Cagiva was also successful in off-road racing, winning the 125cc world motocross title on more than one occasion (a 1987 model seen here), as well as ISDE gold medals and the Paris–Dakar Rally.

(Below) *American Randy Mamola (left) signed for the Cagiva road race team for the 1987 season. He was followed later by another American, multiple world 500cc champion Eddie Lawson. The latter gave the marque its first GP victory in 1992.*

the engines, but the motorcycles themselves would be marketed and sold by Cagiva.

Strangely, it had been Ducati, rather than Cagiva, which had made the first approach, or, more precisely, its owners, the state-controlled VM Group. Ducati's problem at the time was lack of demand, with large areas of its production facilities at Borgo Panigale, Bologna, standing idle. The situation at Cagiva's Varese plant was precisely the opposite.

With this background, it might have been imagined that the June 1983 agreement would have been the platform for a harmonious relationship. In practice, the opposite was the case. The bone of contention was Ducati's refusal to quit building their own bikes, which really upset their partners in Varese. For almost two years there was an uneasy peace. Cagiva bought batches of engines, while Ducati continued building bikes. Press stories concerning Ducati's forthcoming extinction probably did not help.

The Castiglionis may well have helped fuel the Ducati's closure stories for their own benefit. The fact is that Cagiva moved the game forward by reaching another agreement with Ducati's government controller

(probably without the knowledge of Ducati's Bologna staff), and bought the name and plant outright. This was announced to the public on 1 May 1985, and the control of Ducati, lock, stock and barrel, passed from the Italian state to private hands. With this agreement, a new era in Italian motorcycling history began.

Initially, the brothers intended to retain the Ducati name for a short period only, as they had done earlier with HD Cagiva, but they soon realized that this would not be a sound financial move. Quite simply, the Ducati name was too valuable to be cast aside.

A COMPREHENSIVE RANGE

The Castiglioni brothers now had a comprehensive range, from 125 to 1000cc, but felt they still needed one more thing – style. Again, a bold move was made. They acquired the services of no less than Massimo Tamburini, co-founder of Bimota, and its chief designer for over a decade. At first, Tamburini worked from his home on the Adriatic coast. By the late 1980s, he had

been installed in a brand-new design centre in San Marino, known at the CRC (Cagiva Research Centre). This organization to this day is a key facet in the successes of the Cagiva organization and its associate companies, which have included not only Cagiva, but also Ducati, Husqvarna (acquired in 1986), Moto Morini (acquired a year later in 1987) and, later, MV Agusta itself.

For several years after purchasing the name from the Agusta Group, Cagiva simply

(Above) *Another field in which Cagiva has a strong presence is military motorcycles.*

Styled by the F4's designer Massimo Tamburini, the 125 Freccia (Arrow) was one of the top-selling 125 sportsters on the important Italian domestic market during the late 1980s.

(Above) *The Freccia (Arrow) was replaced by the even quicker Mito at the beginning of the 1990s.*

The mighty 900 Elefant used one of the two-valve-per-cylinder Ducati V-twin engines. This is the 1991 version.

banked the MV title and, although many Ducatis were designed and built, together with smaller quantities of Husqvarnas and Morinis, no one saw a motorcycle with the legendary MV Agusta logo. However, Claudio Castiglioni knew the value of the name and wisely bided his time; he let Ducati, in particular, generate the funds needed to re-launch MV in the style it deserved.

The whole Cagiva group came close to financial disaster in the mid 1990s, but the sale of Ducati pulled things around; as far as motorcycles were concerned, the Castiglionis retained all the other marques. Perhaps more importantly in the long term, they retained the services of Massimo Tamburini and the CRC team. In fact, it was Tamburini and CRC who had been responsible for the best of the revitalized Ducati range, including the Paso and 916. The same team was also responsible for much of the later success achieved by the Cagiva 500 GP bikes, and also for the F4 project, which finally made its debut as an MV Agusta in 1998 to worldwide acclaim (*see* Chapter 14).

(Above) *Advertisement for the new Mito Evo, which debuted in 1994.*

The author (second left), after his Mick Walker Racing teamster Dean Johnson (in leathers) had just won the 1995 British Super Teen title on a 125 Mito Evo. Keith Davies (third right) and Richard Davies (far right) are from British Cagiva importers, Three Cross (who are also the official British MV distributors for the new F4 series).

14 F4

RUMOURS AND REALITY

The MV Agusta F4 has been acknowledged by virtually everyone who has been lucky enough to ride one as the best motorcycle of the late twentieth century. In reality, however, it might never have been marketed as an MV at all. Instead, it could just as well have been a Cagiva, a Ducati, or even a Ferrari! The fact that it is an MV represents a fitting end to one of the strangest true-life motorcycling projects of all time.

The F4 saga is really the result of dinner-time discussions between two men – Claudio Castiglioni and Massimo Tamburini – with the added intervention of Piero Ferrari, son of Enzo Ferrari, and Ferrari's owners, Fiat.

First rumours concerning a brand-new Cagiva-masterminded Superbike came at the beginning of the 1990s. Photographs of the new engine, an across-the-frame four, were first released at a Ferrari press conference in 1993 – by mistake. The Cagiva Group had kept the engine behind closed doors, even keeping it out of sight when a fully finished prototype was pictured on the road in Italy. However, when Ferrari Engineering featured a colour picture of the power unit at the launch of its new 465 GT car, Cagiva boss Claudio Castiglioni was forced to confirm that the engine was being developed in conjunction with Ferrari. At the same time, both Claudio Castiglioni and Piero Ferrari had to admit that they had test-ridden the machine.

Castiglioni said at the time of the 1993 press conference that producing target power was 'no problem'. The photograph revealed that the inline engine, topped by a bank of fuel injectors, was similar to half a Ferrari V8. It featured several interesting innovations (*see* the technical bulletin on pages 182–5), including the radial valve cylinder head and cassette-type gearbox.

There is no doubt that Cagiva's financial problems of the mid-1990s held up development of the F4 project – as it did to the

A predecessor to the MV Agusta engine, the 4-cylinder Cagiva unit being fitted up for test in the experimental section of the Ducati factory, 1991.

The Idea and the Legend:
Words from the Boss, Claudio Castiglioni

One evening while discussing with my colleague and friend Massimo Tamburini the topic that characterizes nearly all of our encounters (motorcycles, naturally), the idea came about that was to give life to the F4 project.

Together we imagined an engine for a large-displacement motorcycle, a four-cylinder engine that would give top performance. The project was a complicated and difficult one, and, pressed by our respective commitments, we put it aside for the time being, though promising to take up the subject again soon.

The idea of the engine kept coming back, until the confidence in the technical quality achieved by our projects and the experience gained in the area of development permitted us to consider this idea an actual possibility.

Among the individuals who were at my side during this arduous undertaking, I must mention Piero Ferrari, whose help was invaluable and of such fundamental importance to the realization of the F4.

Naturally, the first step towards seeing the F4 project materialize coincided with Cagiva's acquisition of the glorious MV Agusta trademark, which possesses the most fascinating history in world motorcycling. Its victories on the race tracks of the world generated an indefinite number of enthusiasts throughout the globe, which no one has ever succeeded in equalling.

Seventy-five rider and make World Championships, 270 Grand Prix victories and a total of 3,027 victories in the various different biking disciplines have made the MV Agusta make a particular legendary myth.

We were then to make what was unquestionably the most difficult decision for the achievement of the project: to abandon world racing competitions in the 500 class – the most difficult in terms of technological effort and the most important in terms of stature, as well as an area that saw Cagiva as a major player and one of the leading performers.

This was a strategic decision stemming from a precise intention to devote our specific knowledge and best men to a project that would stimulate the enthusiasm of everyone involved, who were excited just hearing the famed MV Agusta mentioned and eager to return it to its former fame.

The Experimental Department in Varese was assigned the design and development of the engine, while the Cagiva Research Centre, directed by Massimo Tamburini, was given responsibility to create all the other components of the bike, in addition to its styling. The result of this joint effort is a product which will surely provoke comment for its exclusive features and sophisticated mechanics.

The 'radial' valve timing system, a choice inspired directly from the F1, and the special removable gearbox, are clear examples of the commitment to create a racing-inspired engine that is completely innovative.

The technical qualities, sportsmanship and competitiveness fused into an innovative design were the sole principles to guide us in our goal to give all enthusiasts and admirers of the MV Agusta trademark a product worthy of their expectations.

Through difficult, yet passionate work, we achieved the highest goal we had ever set for ourselves. The trademark cherished in the memories of all motorcycle enthusiasts has been revived, and, with the F4, Cagiva has breathed new life into the legendary MV Agusta.

Ducati 916. However, the subsequent sale of Ducati also meant that Cagiva was able to build the F4 without worrying about its effect on sales of the Bologna V-twin series. They could also consider taking the F4 racing. While Cagiva still owned Ducati, the F4 was badged as a Cagiva (but only during prototype development). By the time the motorcycle was finally ready, in early 1998, it was to be called an MV.

The biggest problem in 1998/99 seemed to be putting the design into series production, perhaps because of the Fiat/Ferrari question. Rumours surfaced at various times to indicate that Cagiva would have to pay some sort of fee or royalties to Ferrari, because of the Ferrari design content of the F4's engine unit.

Another piece in the F4 jigsaw – at present unsolved – is the matter of other engine sizes. Besides the current 750, there are also reports that both 900 and 600cc versions will follow in due course.

Whatever the outcome, no one can dispute the fact that the F4 is a superbly engineered and a stunningly styled motorcycle. Not only is it ready for the twenty-first century, but it is also a credit to the MV Agusta marque logos on its red and silver bodywork.

Official factory studio photograph of the F4 Serie Oro, *1998.*

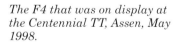

The F4 that was on display at the Centennial TT, Assen, May 1998.

The Cagiva Research Centre has had the privilege of working on a project of great importance and whose worth transcends the mere significance of a sport vehicle to which an engineer would normally make reference.

The F4 has been a project involving a much deeper content, and I feel it my duty here to thank the President Claudio Castiglioni, who gave me the opportunity to dedicate all the resources of the Centre under my direction for the production of a vehicle destined to mark the rebirth of a make, of which I have been a passionate fan.

The F4 for the rebirth of the MV Agusta: this idea immediately had a special significance for me, because the first bike to put me to the test as an engineer was an MV at the time of epic race-track duels of the silver and red bikes of Cascina Costa.

Today, to have contributed to the creation of a complete motorcycle – incorporating an idea shared with the President to initiate a project for a new engine – is wholly satisfying, not only in strictly professional terms, but also because of a passion for the world in which the MV Agusta is a cornerstone.

The concepts that my colleagues at the CRC and I have put into the F4 project are the result of years of activity in this field, but they are, above all, the work of persons that have been skilful in carrying out their work with total commitment and admirable dedication.

The team spirit that can be felt in CRC and the irrepressible will to perform one's work to the fullest potential that permeates each and every department cannot help but have a result such as the F4 project.

I like to think that the special location of the Centre has in some way contributed to the success of this project. The opportunity – certainly not a common one – of working in an environment that has been specially set aside and organized has undoubtedly aided in the success of our research.

I am certain that the F4, for all its technical content, for the technological choices adopted, for the general quality achieved and for its 'personal' design can be considered a vehicle *par excellence*.

I also think that the statement, 'beauty is objective', if applied to the F4, can provide a fertile ground for discussion.

Certainly, this motorcycle fully responds to the principles of 'form follows function' that we had set for ourselves at the start of the study.

I wish to thank all my colleagues at the CRC who worked at my side during this undertaking and Claudio Castiglioni who gave us the opportunity to make it a reality.

I also imagine with great satisfaction the lucky enthusiasts who will be able to personally discover all the special features this high-performance motorcycle of such esteemed value has to offer.

THE TECHNOLOGY

Chassis

Frame

The F4's frame is made up of a 'mixed' steel and aluminium structure. The main section is composed of a trellis in round, chrome molybdenum steel tubes that wrap round the engine, connected at the rear to light-alloy plates that provide a pivot point for the swing arm. It is of a modern design, giving great rigidity.

Cagiva Research Centre engineers have provided this component with high bending and torsional rigidity, low weight, and maximum mechanical accessibility: They have succeeded in continuing the tradition of the past MV 'jewels', while adding modernity with the beautiful light-alloy swing-arm pivot-point plates.

One special feature characterizes the 'mixed' frame of the F4. The trellis/plate connecting points can be separated, to allow the division of the bike into two distinct sections: the front end with the steel cage connected to it, and the rear part with the

Single-side swing arm with five-star wheel.

(Left) *Two buttons for adjustment of ride height.*

(Right) *Single-sided swinging arm. Note also the sprocket and rear disc, and the carbon-fibre chainguard.*

(Below) *Twin allen keys clamp the swinging arm, as shown.*

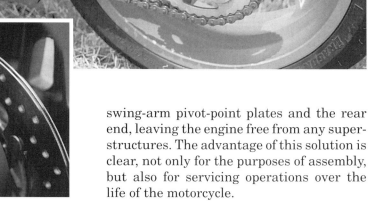

swing-arm pivot-point plates and the rear end, leaving the engine free from any superstructures. The advantage of this solution is clear, not only for the purposes of assembly, but also for servicing operations over the life of the motorcycle.

The trellis is closed at the front on the steering head, which is made of a 'micro-cast' steel structure that also serves as a support to the accessory components of the front end.

Connected to the swing-arm pivot-point plates is the rocker arm of the rear suspension and the removable rear subframe. This trellis is a structure made of round aluminium alloy tubes to which the entire rear section is attached.

This construction philosophy leads to other notable advantages during chain assembly, thanks to the convenience of pre-assembled units.

Swinging Arm

This component plays an important role in motorcycle design, not only because of its main function as a support to the rear-wheel suspension unit, but also because of the visual impact that it gives to a motorcycle.

CRC engineers invested a significant amount of time and technical effort in the research of the F4 single-sided swing arm. Cast in a light alloy, the high level of technical content in this part was the subject of

The CRC logo, representing the Centro Ricerche Cagiva (Cagiva Research Centre).

a university thesis and will surely be further elaborated in high-profile international scientific magazines in the near future. The values of torsional and bending rigidity, the low weight, the highly resistant section of the curved arm, whose innovative 'arc' with integrated truss gives the component uncommon individual style, helping provide the F4 with its unique style.

Structural calculation software called FEM (Finite Element Modelling) enabled very high values of rigidity and mechanical resistance through more than 80 analyses, which reproduced every possible condition the component could encounter.

Equally important was the designers' goal to limit weight as much as possible. And lastly, the care taken in the design of the swinging arm made it into a 'sculpted' object harmoniously integrated into the design mosaic of the F4.

The state-of-the-art design and technology of this component incited CRC to safeguard it from being produced by others through a patent of the 'styling model'.

Suspension

For the suspension, as well as for all other parts of the bike, special components were designed to provide excellent performance.

At the front, the F4 has an upside-down fork with several unique features: 49mm diameter legs for a 113mm wheel travel, generous adjustments for compression, rebound and spring preloading and, above all, a system for fixing the front-wheel spindle with a quick-release clamp. These features all endow this assembly with top-rate performance.

To emphasize how detailed CRC's research was, the fork axle lugs were designed with the ability to be quickly detachable. The fork legs were specially coated to provide a smoother action. The

Front forks, featuring a quickly detachable wheel spindle. The discs are fully floating.

(Below) *Near-vertical rear shock absorber.*

benefits in the functional performance's suspension system are a real bonus.

Completing the fork unit are upper and lower yokes specially designed for the F4. The steering head, cast in an open-sectioned light alloy, has a characteristic diamond-patterned prism design that gives the piece greater stiffness and a rugged appearance. The lower yoke, also cast in light alloy, but with a closed section to support the greater stresses it is subject to, has a 'bridge' shape with clamps conveniently distanced from those of the steering head.

The design of the yoke also acts as the conveyor of air to the radiator.

The rear end features a 'floating' suspension unit based on a single air/oil shock absorber driven by kinetic motion operating with a 'progressive articulation' that ensures rear-wheel travel of 120mm.

The shock absorber, which can be adjusted for compression, rebound and spring preload, is activated by a special rocker that can be placed on two different pivot points,

which allow different 'progression curves'. The rocker in turn is moved by the swinging arm by means of a special counteracting structural element with a variable length. This allows, among other things, the bike's set-up to be varied.

To grant this system the utmost operating performance, this special element is mounted with ball and socket bearings with steel/PTFE contact surfaces giving a low friction coefficient.

Steering-Head Angle Adjustment Device

Profiting from concepts successfully implemented on previous Tamburini designs, CRC engineers fitted a device on the F4 designed to adjust the steering-head angle. This device gives the possibility of regulating the important parameter of the trail, while keeping the wheelbase unchanged.

This is achieved through an eccentric support of the steering bearings, of which the seats are misaligned with respect to the theoretical axis. With a simple adjustment, different steering-head angles are obtained to conform to the personal specification of professional rider and road motorcyclists alike.

This device used on the F4 was the focus of in-depth theoretical analyses. The results, in comparison with other laboratory tests, led to the definition of a new system for locking the steering-bearing eccentric. The feature has also been patented by the CRC.

Steering Damper

Clearly derived from racing models, the steering damper is located transversally in relation to the forward direction of the vehicle. This is to allow the greatest performance with perfectly symmetrical responses for left and right movement of the steering unit.

The F4's steering damper is anchored to the frame at both ends and is activated by means of a connecting rod attached to the steering head. The system thus allows optimum operation of the 'linear' hydraulic unit used, while ensuring the elimination of

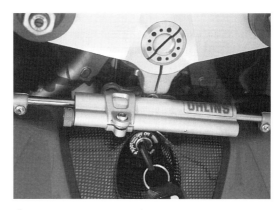

Öhlins hydraulic steering damper mounted across the frame as on Ducati 916 series (and Cagiva Mito Evo) – a Tamburini trademark.

transverse factors of forces and the resulting friction induced by them on the system. The device is optimally integrated into the motorcycle, being mounted in a protected position that gives total access, and allows adjustments, even while the bike is in motion. The innovative solution has also been patented by the CRC.

Wheels

Special focus was given to the design and function of the wheels. The limitation of certain important parameters, such as weight and the effects of inertia, have been overcome by a new 'star' design.

The F4's front has a state-of-the-art 3.50 × 17in rim with a five-pointed 'star' design embellished with thin elliptical spokes connected to a hollow pentagonal hub. The special design gives the whole front end a light, clean visual impact.

The superb attention to these details is topped off by the hollow wheel spindle, which has a thin 35mm-diameter wall offering extreme rigidity and a low weight.

At the rear, a massive rim measuring 6.00 × 17in has been chosen. Although its form echoes the pentagonal star shape of the front, its design is deliberately different from its twin at the front.

Attachment to the rear-wheel spindle is the job of one large central nut that is similar in terms of its functional concept and size to the one used for all the current F1 racing cars.

The simplicity of use of the rear-wheel attachment system, in conjunction with the single-sided swinging arm, makes the solution appreciated not only during the assembly phase, but especially during servicing.

The rear-wheel spindle, constructed from a large forged billet, machined to a high-quality tubular structure, 50mm in diameter, is complemented by the attachment of the final-drive rear sprocket fitted with a 'multi-lobed' radial rubber cush drive.

The whole unit is supported by an eccentric hub that offers the means of adjusting the chain tension.

The tyres were specially designed for the F4. The CRC engineers, together with the technicians from the supplying companies, opted for a very particular size for the front tyre of 120/65 ZR 17, whereas the rear is 190/50 ZR 17.

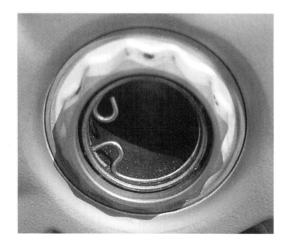

Circlip for quickly detachable rear wheel (there is also a large nut).

Braking System

The importance of the F4 project, and the desired homogeneity, led to every component of this system to be purpose-designed for the F4.

The front brake consists of calipers, each of which has six opposed pistons of different diameter, derived from Cagiva's racing expertise. Special attention was paid during the design stage, giving an optimization of the internal passages, to guarantee an excellent 'bleeding' of the system. The master cylinder used for these calipers is a masterpiece of technology and design. No less than three different industrial patents were needed to protect the new ideas that were put into practice.

Technology, performance, quality and design were the watchwords guiding the designers throughout this project. The peculiar layout of the master cylinder, original and dimensionally reduced brake-fluid tank, and the lever shape (incorporating span adjustment and predetermined braking point) exemplify this approach. The same approach was applied to the design of the clutch lever assembly, giving excellent visibility of the instruments.

Two 310mm-diameter floating discs held by flanges specially designed to merge with the wheel complete the front brake set-up.

The rear boasts a four-pistoned caliper, which is claimed to out-perform all others with similar characteristics that are currently available on the market. The pads give 30 per cent more braking surface and 49 per cent more material volume, with obvious advantages in terms of the performance and life of the component.

A single 210mm disc, with reduced width of the braking surface, reduces the unsprung weight and gives a visual effect of extreme 'lightness'.

(Above) *The twin stacked headlamps allow a smaller front area for the fairing.*

Lights

How could such an important part on the F4 fail to meet the expectations of the most demanding enthusiasts?

The CRC stylists and engineers came up with an innovative front light. The twin stacked polyellipsoidals give an unmistakable character while also allowing the motorcycle to be homologated worldwide, reducing weight and rendering possible a very narrow fairing for maximum aerodynamic (small frontal area) and styling advantages. This, as for the many other new ideas used on the F4, has been protected by a patent.

A multifunctional die-cast support holds the lights, instrument console, fixing points for the side fairings, air-intake ducts and two arms holding the headlight fairing and mirrors. The result is a really 'clean' layout that makes the cockpit and the front end of the bike appear perfectly integrated.

The rear-view mirrors are also integrated into the highly profiled headlight fairing, and incorporate the front indicator, thus maintaining the overall clean styling of the front end.

Another clever styling feature – the front direction indicator built into the mirrors.

Rider's-eye view of the F4 cockpit, showing LCD displays and warning lights. The white-faced Veglia tachometer not only looks the part, but is extremely easy to read.

(Below) *Thumbwheel adjuster for the clutch lever; operation is hydraulic.*

At the rear there is a twin-lamp light with triangular lenses set into the seat-tail's end section.

Instruments

The bike features a high-tech instrument console of original design. The 'mixed' composition of the instruments includes an 'analogue' electronic rev counter, a multifunctional display and a set of warning lights. The rev counter, redlined at 17,000rpm, has a large dial to ensure legibility even at high speeds.

The display oversees the tachometer, odometer, double trip, water temperature and clock. The parameters can be set in kilometres per hour and Celsius degrees, or miles per hour and Fahrenheit degrees.

The warning lamps signal the lights, high beam, neutral, direction indicators, side-stand, current generator, rev limiter and oil pressure. The components of the instrument console are enveloped in a stylishly designed case that facilitates placement in the aerodynamic form of the headlight fairing without obstructing visibility.

Controls

The F4's manual controls include light-alloy semi-handlebars anchored into fork sleeves by means of practical 'bracelet' clamps. Several control accessories are mounted on the handlebars through practical pins that guarantee the reciprocal layout, which was determined after detailed ergonomic studies.

The optimum slant, the perfect grip, and the ergonomic levers (which are adjustable)

Right-hand switchgear and twin throttle cables. Note the small choke lever on the throttle body.

(Below) *Aluminium prop stand, which folds out from the fairing as shown.*

give the F4's cockpit absolute efficiency and operative ease.

As for the pedal controls, the bike is equipped with supports manufactured in aluminium alloy that have a patented feature of adjusting the set-up of the foot pegs through a highly practical eccentric system. The gear change and brake levers are also fitted with an adjustment system providing the optimum layout with respect to the foot pegs. The foot pegs, in aluminium alloy, have a specially formed block pattern; this makes it easy to shift the position of the foot, while offering a secure and efficient foot grip.

Styling

Achieving the look of the F4 was not a matter of mere aesthetics.

Studying the styling of a vehicle, especially of such a high-performance bike as the F4, is one of the most complex parts of the process that the designers must face. The work of Massimo Tamburini and his CRC staff, the F4's styling is as performance-driven as is possible for an 'object' designed to meet the high expectations of demanding users and biking experts.

Tamburini describes the body as:

A mosaic of parts connected by racing-inspired quick fasteners. It is the result of the excellent integration of the company's engineering and design departments. The reciprocity of the stylistic/functional demands compared with those of the frame and engine, led to a synergy among those departments that yielded stunning achievements. The development of the style closely followed the project of the entire frame, and led to a sharp, elegant exterior with overall dimensions comparable to those of a motorcycle of much smaller displacement.

The effect is a streamlined body that blends the best aerodynamic performance and awe-inspiring looks derived from racing models, and a compactness that only the co-ordinated research of chassis and engine designers could produce.

The synergy between different functions gave the opportunity to pinpoint every single detail, even those apparently insignificant, to ensure an optimum integration of the three important sections: the body, the chassis and the engine.

A special focus on the bike's ergonomy concluded in a riding position that gives the rider a feeling of complete integration with the bike, making riding the F4 into a concentrated dose of performance-induced adrenaline.

The aggressive look of the pinpointed headlight fairing, which enfolds and extends out the front, almost to 'swallow' the front wheel, the handsome as well as purposeful tank, and the single-seater tail to top off the rear section, which is graced with an exhaust system of rare beauty, makes the F4 a milestone of style and technology.

The Rider: Words from Giacomo Agostini

MV Agusta is part of my life. It represents for me a real symbol of many joys, hopes and great triumphs. I have to share my greatest victories with this mythical make: as a matter of fact, thirteen of my fifteen World Championship titles were won with an MV Agusta motorcycle.

During the years, MV Agusta has become popular all over the world among the sportsmen and sporting enthusiasts and has been able to incite admiration and great passion in conjunction with its name. A real enthusiasm and an immeasurable competitive passion sustained the activity of the Agusta family, through a personal involvement.

The same interest and the same passion explains Cagiva's choice of giving new life to the mythical MV, the trademark which has influenced most of the world motorcycling history.

The MV Agusta F4, besides a name famous and known worldwide, can boast advanced technological research and high-quality mechanical innovations.

I will be really moved and proud seeing again on the race tracks and roads the Italian motorcycle with its characteristic logo: the one of the inimitable MV Agusta.

Tamburini-styled single racing seat.

Individual number of the test machine: Number 28 of 300 (Serie Oro).

LIST OF CRC PATENTS

Chassis

- Front headlight including stacked polyellipsoidals.
- Tapered locking device for the steering-bearing seat element.
- Steering damper located transversally in relation to the forward direction of the vehicle.

Massimo Tamburini – A Design Genius

Now 56 years of age, Massimo Tamburini is today rightly considered a genius in the field of motorcycle design. The creator of the stunning MV F4 can be proud of his record over the last three decades.

Tamburini first sprang to fame in 1968 when he converted several 600 MV four-cylinder roadsters into stunning sports bikes. In 1966, he had set up a heating systems business with two friends named Bianchi and Morri; seven years later, these same three founded Bimota Meccanica, and went on to produce many innovations on racing and street bikes. Johnny Cecotto won the 1975 350cc world title on a Bimota-framed TZ Yamaha-engined device, and Randy Mamola also began his GP career on just such a machine, and these were followed by class-leading Superbikes such as the SB2, KB1 and KB2.

After quitting Bimota in the early 1980s, Tamburini was responsible first for the Roberto Gallina 500cc GP Suzuki effort. He then joined Cagiva in the mid-1980s, reshaping the previously unsuccessful 500 four-cylinder two-stroke into a winner.

Financed by the Castiglioni family, Tamburini set up the Centro Ricerche Cagiva (Cagiva Research Centre), more commonly known as CRC, in a hill-top retreat in San Marino. With some thirty-five staff, including his son and one of his two daughters, Tamburini has been responsible for creating such notable street bikes as the Cagiva Mito 125, Ducati Paso (named after his boyhood hero Renzo Pasolini, who died at Monza, together with Jarno Saarinen, on 20 May 1973), Ducati 916 and, most recently, the awesome MV Agusta F4.

Growing up just outside the Adriatic resort of Rimini, Massimo Tamburini always had motorcycles around him. His father rode a large-capacity Moto Guzzi, and he witnessed at first hand the local street races where stars such as Agostini, Hailwood and Pasolini took part.

In a 1998 interview with journalist Roland Brown, Massimo Tamburini had this to say about the new MV four:

The F4 is the most satisfying – the one I like the best.

With most bikes you start with something from a previous model, but on the F4 we did not have to make any compromises. Mr Castiglioni [Claudio] said we could do what we wanted. To start with a clean sheet of paper and a legendary name like MV Agusta is the perfect situation for an engineer. We studied all the components totally afresh, from a viewpoint of both function and style. It was a fantastic opportunity – and we are very happy with the result.

Roland Brown describes Tamburini as 'perhaps the most influential figure in modern motorcycle design'.

- Front brake/clutch lever unit with multiple safety system.
- Front brake/clutch pump unit lowered to reduce dimensions.
- Front brake/clutch pump tank with safety quick mount.
- Multiple tank unit with 'tandem' layout of components.
- Adjustable foot-peg support.
- Rear single-sided swing arm (inventor's patent for styling model).
- Exhaust silencer (inventor's patent for styling model).

Engine

- Multiple-sectioned duct with special geometry for intake and exhaust systems of internal combustion engines.

THE ENGINE

For the first time in Italy, designers have made possible a mass-produced four-cylinder motorcycle engine that can fearlessly compete with the best of the world's competition. The F4's arrival in production, and its achievement in terms of reliability and performance, thanks to unprecedented technical solutions for large production bikes (such as the cassette gearbox, and radial valves with direct control), reward the efforts of the engineers who have met such an important challenge.

General Layout

The four cylinders of the MV Agusta F4 represent the standard for comparison in the Supersport category. This is the result of the interaction and collaboration of the various engineers assigned to the vehicle and the engine. Never before in a mass-produced bike have the needs for functional reciproca-tion among the different parts of the frame and the engine been taken into such great consideration by the different work groups.

The shapes of the fairing have been created to be in harmony with the small pick-up cover and the special double-slanted cut (the only one in motorcycle design) of the clutch cover, which aids in reducing the size of the lower part of the engine, while giving it a totally unique aesthetic effect.

The size of the cylinder head and its cover have made it possible to reduce the dimensions of the fuel tank to a level never seen before with a four-cylinder.

Removable Gearbox

A removable (cassette-type) gearbox has been adopted for the first time on a mass-produced four-cylinder, following the experience of the Cagiva Racing Department during the years of the 500 World Championship.

Due to the availability of a complete set of specially designed gears, it will be possible for both the road user, and the racer at all levels, to optimize the gear ratios of the motorcycle, depending on the track or the average usage conditions.

The Radial Valve Cylinder Head

For the first time on a high-performance road bike, a timing system has been adopted with four radial valves per cylinder, with direct control of the valves by means of tappets and tapered cams. The radial structure of the valves allows them to move in such a way as to distance the head of the valves from the wall of the cylinder with the increase of the lift. Conditions of fluxing and filling are thus created that are considerably improved over the traditional parallel valves.

The increased fluxing, and especially the shape of the flow inside the cylinder, have been analysed using the same equipment and technologies that were used in the quest for the highest performance of the Cagiva 500 (which nearly captured second place in the 1994 World Championship). This research has resulted in a perfect harmony of torque and horsepower. The motorbike has an unmistakable character, thanks to a smooth supply of power, without the troughs, and is guaranteed the greatest performance possible.

The Timing System with Transmission

Nothing has been spared in the search for performance. In order to attain a highly compact combustion chamber, and the smallest angle between the valves in the class (22 degrees), without running up against a decrease in the resistance of the middle zone of the driving shaft, a specially designed transmission system was adopted to control the timing chain. In this way, the toothed wheels on the camshafts could be reduced in diameter, allowing for the most compact head and cover unit in the class.

Fuel-Supply Intake System

The F4 uses a sophisticated fuel-supply intake system, on which the most valuable component is the electronic injection. This utilizes a special processing unit that grants the bike optimum fuel-supply performance, thus giving reduced consumption and emissions. The system also includes two intake silencers with dynamic air vents that open out towards the front of the bike in the pressurized air zone. These silencers convey air to the 'exposed' air box, which has a larger volume to allow the engine to 'breathe' better.

The function of the dynamic air vents is optimized by lateral slots that perform the function of 'guiding' the flow towards the air box, thereby reducing the vortexes that pre-

Aircraft quick-action fuel filler complete with MV logo.

vent air from filling the air box and creating a depression effect in the direction of the flow.

The air box is mounted on top of the throttle bodies in an original 'sequential' layout with the fuel tank and the cooling liquid expansion tank (a patented feature). It has been designed to ensure maximum capacity while harmoniously completing the styling of the whole area on top of the engine.

Cooling System

This part of the project defined by the CRC also includes top-grade components.

The large concave radiator matches up to the shape of the body sides, closing the front section fairing to give optimum efficiency. This component has been conceived to obtain a high number of 'elements' and has been equipped with a system of 'high-efficiency turbolators', which provide a high level of thermal exchange. The radiator is assisted by an electric fan that is activated by means of a specific flow conveyor.

To complete the cooling system, a 'heat exchanger' is connected to the cooling-liquid radiator. This enables the engine's oil temperature to be maintained at ideal levels.

Exhaust System

One of the most prized components that further adds to the great technical/stylistic character of the F4 is the exhaust system. For this, CRC engineers devised a technical solution that goes beyond the confines of mere 'function'.

The four pipes exiting the engine take a sinuous course, enveloping the engine,

(Above) *Tamburini trademark exhaust, a progression of the type conceived for the Ducati 916 series.*

(Right) *Pasini-built F4 Mini Project. The specification includes mechanically operated front and rear disc brakes, three-spoke wheels, slick tyres and a single-cylinder two-stroke engine with automatic transmission. Stafford Classic Bike Show, 20 May 1999.*

perfectly copying its forms, to lead to a totally original expansion/silencing system.

The system is composed of symmetrical 'elements', in which the two round expansion units, specially fitted with a diaphragm, open out by means of silencer pipes with variable lengths. The 'organ-pipe' system allows the noise level to be significantly decreased, while giving the system highly personalized 'vocals' that have been 'tuned' on the basis of musically derived algorithms.

Poster showing Agostini and the F4 celebrating the Mallory Park Post-TT-Bike Fest, June 1999.

F4 *Serie Oro* (2000)

Engine

Type	Liquid-cooled dohc
Bore and stroke	73.8×43.8mm
Capacity	749.4cc
Compression ratio	12:1
Carburation	Electronic fuel injection
Lubrication	Wet sump, gear pump
Max. power (at crank)	126bhp @ 12,200rpm
Maximum torque	7.3kgm @ 9,000rpm (53ft/lb = 72 Nm)
Fuel tank capacity	4.8 gallons (22 litres)

Transmission

Gearbox	Six speeds, constant mesh, cassette-type
Clutch	Wet, multi-plate
Primary drive	Gear
Final drive	Chain
Ignition	Weber-Marelli 1.6 M ignition/injection system with induction discharge electronic ignition
Fuel system	As above

Frame

Modular construction with rear swinging arm in magnesium alloy (aluminium alloy for F4S). Chrome-moly steel tubular trellis frame with patented adjustable steering system

Suspension and steering

Suspension	front	49mm inverted hydraulic front fork with quick release for front wheel
	rear	Single shock absorber with adjustable rebound, compression damping and spring preload

Wheels

		Five-spoke magnesium alloy wheel (aluminium alloy for the F4S)
Tyres	front	120/65 ZR17
	rear	190/60 ZR17

Brakes

	front	Twin six-piston calipers with double 310mm steel floating discs on an aluminium flange (steel on F4S)
	rear	Single four-piston caliper with 210mm steel disc

Dimensions

Length	80in (2,026mm)
Width	27in (685mm)
Seat height	31.5in (790mm)
Ground clearance	5.12in (130mm)
Dry weight	397lb (180kg)

Performance

Top speed	171mph (275km/h)

The placement of the silencing system in a protected area beneath the seat, its functional capacity and the expression of the technical/design concepts (patented for the 'styling model') make this component unique.

The Art of the Motorcycle

From June to September 1998, the Solomon R. Guggenheim Museum, on Fifth Avenue, New York, staged its most popular exhibition ever – the Art of the Motorcycle – attracting over half a million visitors.

Spanning over one hundred years of the motorcycle, the display included ninety-five machines, the oldest being the 1894 Hildebrand and Wolfmüller and the newest, the 1998 MV Agusta F4. The former needed 1489cc to generate 2.5bhp at 240rpm and give a top speed of 28mph (45km/h), while the latter needed only half the engine displacement of 749cc to pump out 126bhp at 12,200rpm and attain a maximum speed of 171mph (275km/h). With that rate of progress, what will the next one-hundred years bring?

Besides the new F4, the other MVs chosen for display included a 1956 500 four-cylinder GP bike and a 1973 750S sportster.

'The Art of the Motorcycle' was so successful that it was transferred to the Field Museum of Natural History in Chicago from November 1998, before crossing the Atlantic to the newly constructed Guggenheim Museum in Bilbao, Spain, for 1999.

If the exhibition, which used a mass of gleaming stainless steel to display the bikes, was impressive, so too was the show catalogue, which had a mammoth 432 pages. The photographs, in both colour and black and white, were simply stunning, reproduced using printing techniques and art paper of the highest quality.

MV's brand-new F4 could hardly have had a more high-profile launch than a presence as the star of the most prestigious motorcycle show the art world has ever seen.

Testing the F4, by Mick Walker

In the spring of 1999, I became the first person, outside of the staff of British importers Three Cross, to ride the new F4 on UK roads. This came about because of my past involvement as race-team manager for Three Cross, and having its boss Keith Davies as a personal friend of over twenty-five years' standing.

I had already viewed both the limited production and ultra-expensive F4 *Serie Oro* and the series production F4S at shows both in the UK and abroad, so knew what to expect visually – or so I thought.

However, seeing the bike – a *Serie Oro* in the flesh so to speak – outside a show hall was an altogether different experience. My first thoughts were that it was a truly magnificent piece of engineering. Quite simply, it makes every other street-legal motorcycle seem old hat – even the Ducati 996 Fogarty Replica, which I had tested earlier.

But what would it be like to ride? I was fully expecting this to be something of a let-down after the visual experience, but no, I was not to be disappointed. The F4 is as good in practical terms as it is from a styling and technical viewpoint.

What strikes you straight away is this bike has been designed and built by motorcyclists as everything is in exactly the right position and it all works. Accelerating away from rest, the engine strikes the heart for its smoothness, useful power and the noise it makes – and the gearbox is truly superb, a hot knife through butter would be an apt description. Touch the brakes and they are powerful yet exceedingly safe in their operation. The handling and road-holding too give you confidence. You feel you could actually fit racing tyres, tape up the lights, remove the side stand and go racing, it's that good. I also found myself downshifting the gears just to hear the glorious howl from the free-revving engine.

But the biggest surprise was just how flexible the engine was. I was fully expecting it to need constant gear changing. But no. It's

The author takes a break in the Dorset countryside whilst testing the F4 Serie Oro, *June 1999.*

possible to drop down to 30mph in top (sixth) and accelerate cleanly away (full throttle at this low speed is not advised!).

Most of the day-long test session took place in Dorset and on minor 'B' roads at that. The MV is quite at home in this environment. Even at low speeds, the riding position is comfortable. Obviously it does not have the armchair ride of a Honda Gold Wing or even a CBR1000, but by single-purpose sports bike standards it is comfortable.

All the ancillary equipment, including the instrumentation, switchgear and electrical equipment, seem to be top class – an area where past Italian sports bikes have often been weak.

As for attention to detail, this can only be described as breath-taking. For example, it took photographer Vic Bates and myself at least three times longer than usual to take in all the features of the F4. The quickly detachable front and rear wheels, the sculptured lines, the superb alloy castings, the curvaceous exhaust, the star-like wheels, the ultra-clean and precise welding of the steel frame; there's even a couple of press buttons on the swinging arm to adjust the ride height! Perhaps the only area where I thought an obvious improvement could be made was that I was always worried that the side stand (there is no main stand) might dig in and the F4 fall off its perch. As I was riding the £25,000 limited-production F4 *Serie Oro*, damage was always uppermost in my mind (not helped by a 'bring it back as you took it' warning from Keith Davies).

Compared with the standard (regular) F4S, the *Serie Oro* has a number of changes: magnesium alloy (instead of aluminium alloy) for the swinging arm and rear subframe assemblies (with a gold instead of grey finish); carbon-fibre bodywork (instead of heavier fibreglass); aluminium floating disc flanges for the front brakes (instead of steel); and magnesium wheels (instead of aluminium alloy).

There are some other less noticeable differences, such as to the colour of the front mudguard support and adjustable foot controls. The mechanical specification as regards the engine and suspension packages are the same on both models. So is the near-double price for the *Serie Oro* over the F4S really worth it? The truthful answer is no, but of course someone who can afford £25,000 instead of £13,000 will probably not notice the difference. He will also be able to boast that his MV is identical to the ones owned by Claudio Castiglioni, Giacomo Agostini and King Carlos of Spain.

When writing this test in the summer of 1999, there was also another problem for prospective F4 owners – delivery time, which was already a full year in the UK and getting longer by the day. Well, that old saying that the best things in life are always worth the wait certainly applies to the MV Agusta F4.

Appendix: The MV Agusta Owners' Club of Great Britain

The owners' club was founded twenty years ago by a small band of MV enthusiasts and has grown into a well-supported club, with a membership of around 800 over the years, averaging out to about 200 members at present.

In the early years, the club arranged runs and meets, generally around the Midlands, which seemed central enough for a membership that was spread far and wide. The meetings were usually road tours with some invites from other clubs to their track days. A quarterly club magazine was started and a committee formed, with the annual AGMs taking place in a nearby pub. The AGMs are now held at the National Motorcycle Museum, generally in early May.

The late Peter Eacott was elected as spares person within the club, and all spares were ordered through him direct from the Agusta factory. Over a period of time he amassed important spares such as pistons, clutch plates and the normal service items. When the opportunity came to purchase the last remaining spares from Agusta, some keen members clubbed together and, after lengthy negotiations with Mr Laudi and his superiors, some 10 tonnes were purchased. This boosted the club's spares section to cover models dating from 1947 to 1977. It was quite a task to sort, label and identify all the bits and pieces, but now the club has a good range of spares for most machines. The last of an item is kept, so that further parts can be made.

The list of patrons has grown, and now includes John Surtees MBE, Phil Read MBE, Cecil Sandford, Arturo Magni, Stuart Graham and Bill Lomas, a very famous line-up. Thanks to the support of the patrons, the club has been invited to some exciting events throughout the years, with members able to display their machines and participate at venues such as the Goodwood Festival of Speed, Brands Hatch, Monza, Montlhéry, Vallelunga near Rome, Beaulieu and Cascina Costa, the Agusta family home near Gallarate in Northern Italy.

The club has an excellent relationship with the Italian ex-works mechanics and enthusiasts who live close to the Agusta factory; Arturo Magni, the works team manager for the full duration of MV's racing career, Lucio Castelli, also a works mechanic, as well as Romano Columbo and Ubaldo Elli, who own an impressive collection of MVs. Elli has more than thirty ex-works machines, ranging from single-cylinder racers, twin-cylinder prototype machines, and bikes with three, four and six cylinders.

Until two years ago, the club hired the Grand Prix Circuit at Monza for its main annual track event; the rising cost of hiring the track, and new limits on noise have put a stop to this event. The situation is similar in the UK, at the Cadwell Park circuit, where Brands Hatch Leisure has upped charges, and has had to be stricter on noise levels. The intention is now to hold a twice-yearly event at Montlhéry, near Paris, where it should be possible for club members to ride their fabulous four-cylinder machines unsilenced.

The club has a wide and varied membership from around the world and, with the launch of the new Cagiva/Agusta F4, 'soon to be' owners are now joining up. This will bring some new blood into the club and hopefully they will enjoy the sights and sounds of the older bikes that have given the MV name its reputation in the motorcycle world.

The MVOC now produces its magazine in-house, and, with its website, which has generated much interest, is getting to grips with the twenty-first century.

MV Agusta Owners' Club of Great Britain
61 Grimsdyke Road, Hatch End, Middlesex
HA5 4PP, UK
Tel/fax: 02084 282027
E-mail: mcelectron@aol.com

MV Club members Richard Venables (on bike) and engineer/friend Ralph Lemon at Auto Italia, Brooklands, June 1998.

Index